millipedes
crustaceans
arachnids
ONYCHOPHORA
ARTHROPODA
horseshoe crabs
NEMATODA
PLATYHELMINTHES
cephalopods
ANNELIDA
gastropods
MOLLUSCA

PRAISE FOR *THE MODERN BESTIARY*

"Everyone who loves wildlife—especially fantastically weird and cringingly gross wildlife—should read this masterful book."

Mark Carwardine, author/presenter of *Last Chance to See*

"Even after half a century—and counting—as a professional zoologist, I encountered new and intriguing facts on every page, all conveyed in an easy, friendly style. If ever there was a book that highlighted the bewildering wonders of the natural world, and the need for their conservation, this is it."

Michael Brooke, author of *Far from Land*

"If you love animals, especially ones with unsavory habits, this book is for you. Entries are crafted with affection, cast-iron scholarship, and an unyielding dedication to exposing as much hilarious weirdness as the animal kingdom can offer. And that, it appears, is rather a lot. This is a book to adore."

Tom Moorhouse, author of *Elegy for a River*

"This modern bestiary is a magnificent miscellany that will amuse and amaze. From butterflies that make crocodiles cry and penis-fencing slugs to fish that live inside sea cucumbers' bottoms—the natural world is stranger than you could ever imagine."

George McGavin, entomologist, author, and TV presenter

"Just when you thought the natural world couldn't get any more bizarre, you turn the page and learn about a small marsupial that mates itself to death. There's plenty inside this beautifully written book to make you laugh, squirm, and—perhaps most importantly—appreciate how lucky we are to not have to live inside an anus."

Yussef Rafik, zoologist and wildlife TV presenter

THE MODERN BESTIARY

A Curated Collection of Wondrous Wildlife

Joanna Bagniewska

with illustrations by Jennifer N. R. Smith

Smithsonian Books

Washington, DC

First published in 2022 by
WILDFIRE
an imprint of HEADLINE PUBLISHING GROUP

1

Cataloguing in Publication Data is available from the British Library

UK Hardback ISBN 978 14722 8960 5

Typeset in Baskerville MT by Seagulls.net

HEADLINE PUBLISHING GROUP
An Hachette UK Company
Carmelite House
50 Victoria Embankment
London EC4Y 0DZ

www.headline.co.uk
www.hachette.co.uk

Published in United States of America and its territories, Canada and dependencies, and the Philippine Republic by Smithsonian Books

US ISBN: 978 15883 4730 5

This book may be purchased for educational, business, or sales promotional use. For information, please write:
Special Markets Department, Smithsonian Books,
P.O. Box 37012, MRC 513, Washington, DC 20013

Library of Congress Cataloging-in-Publication Data

Names: Bagniewska, Joanna, author.
Title: The modern bestiary : a curated collection of wondrous wildlife / Joanna Bagniewska ; with illustrations by Jennifer N.R. Smith.
Description: Washington, DC : Smithsonian Books, 2022. | "First published in 2022 by Wildfire, an imprint of Headline Publishing Group"-- title page verso. | Includes bibliographical references and index.
Identifiers: LCCN 2022019552 | ISBN 9781588347305 (hardback)
Subjects: LCSH: Animals--Canada--Miscellanea.
Classification: LCC QL45.2 .B34 2022 | DDC 591.971--dc23/eng/20220525
LC record available at https://lccn.loc.gov/202201955

Printed in the United States of America, not at government expense
26 25 24 23 22 1 2 3 4 5

To the large Aussie mammal,
the precocial small primate and the altricial,
freshly emerged larva: to my family

Contents

Water

Air

Introduction

Before we start, let me warn you that this is not really a book for children—so if you bought it for your baby sister, kid nephew, or little cousin (because it's about animals and it has pictures), you need to be prepared for a, perhaps awkward, conversation about traumatic insemination,[1] matriphagy[2] or apophallation.[3] Animals are gross. And gory. And obscene. And I suspect that it could be difficult to address issues such as siblicide with children who are going through the phase of hating their brother or sister.

Now that that's out of the way, welcome! Welcome to *The Modern Bestiary*!

First of all, what is a bestiary—and why write a modern version of it?

Medieval bestiaries were books of beasts: collections of creatures containing natural history information (factual or otherwise), doused in didactic sauce with a strongly Christian flavor. While animal stories have been popular for centuries, across all cultures, bestiaries of the Christian West were written with a very particular goal in mind. Early Church Fathers, such as the third-century

1 When a male uses a very sharp sex organ to stab directly through the female's abdomen into her ovaries. See Common bed bugs, page 28.

2 Babies eating their mothers. See Pseudoscorpions, page 58.

3 Essentially, the phenomenon of biting off your partner's penis after sex. See Banana slugs, page 18.

scholar Origen of Alexandria, believed that God speaks to people through creation. If we pay close enough attention to the natural world, he claimed, we will be able to grasp a deeper understanding of the Creator—and of human nature. Animals and their behavior therefore presented a perfect opportunity for a moral and theological lesson. Some of these lessons were plausible—for example, ants working together for a common good show us how humans should work in unity. Others, however, were completely off. Famously, the pelican was said to kill its young and then pierce its own breast to revive them by feeding them blood; in doing so, the bird symbolized the love and sacrifice of Jesus Christ. Mundane facts (that pelicans feed on fish and not the blood of their parents) did not get in the way of a good allegory. To further blur the line between fact and fiction, the bestiaries contained fantastical beasts, such as unicorns, phoenixes, or mermaids, alongside the real animals.

The first of these moralizing collections of beasts was written sometime between the second and fourth centuries AD, probably in Alexandria. This book, known as the *Physiologus*, sourced animal stories from Egyptian, Hebrew, and Indian lore, as well as the works of natural philosophers such as Aristotle (*De animalibus*) and Pliny the Elder (*Naturalis Historia*), and used them to explain the Christian doctrine. The *Physiologus*, originally written in Greek, was translated into many European and Middle Eastern languages and became hugely popular across Europe. The original contained fifty-odd chapters describing animals (actual and fantastical), plants and rocks, covering everything from ants to lions to antlions.[4] Over the next centuries, more entries were added—and eventually the Middle Ages saw a real bestiary boom.

The illustrations complementing each entry were an important part of the bestiaries. The stories themselves (with their associated

---

[4] There is such a thing as an antlion (in fact, there are about 2,000 species of these insects from the family Myrmeleontidae), but the one in the *Physiologus* is a mythical beast with the face of a lion and the body of an ant, created from the mating of the two creatures. Unfortunately, it suffers something of a design fault in that the lion part can only digest meat and the ant part can only digest grain, so it always starves to death.

moral lessons) were not new or unknown, having been told and retold in many circumstances; therefore, the purpose of the artwork was predominantly to jog the memory of those who couldn't read. As such, the illustrations were meant to be evocative, expressive, and didactic. Unfortunately, they were not always accurate or even realistic, mainly because the artists did not have much of a chance to see the more exotic species (or none at all when it came to the manticores, griffins, or centaurs). Instead, they based their drawings on previous illustrations, descriptions from earlier books, or other people's accounts, forming a sort of composite image created via Chinese whispers, with any gaps filled by their own poetic license. Consequently, many animals didn't look quite right—with lions having very human-like faces, pelicans sporting sharp, eagle-like beaks, whales and dolphins covered in scales, and so on. And the imagery prevailed in other forms: every time I walk through the center of Oxford and peek into Corpus Christi College, I am greeted by a statue of a stubby-beaked, breast-piercing pelican standing atop a sundial in the main quad.

Because the main function of the bestiaries was allegorical rather than scientific, their authors focused on retelling the didactic stories rather than on fact-checking them from the zoological angle. As a result, some myths have been propagated for centuries, even though the authors surely must have cottoned on to the fact that the swan does not have a beautiful voice or the mouse is not born from soil. In fact, some of these odd ideas still persist today, such as the notion that hedgehogs carry apples on their spines. This myth presumably came from hedgehogs being spotted in orchards, snuffling around fallen fruit in search of invertebrates. Some authors went further, describing how these animals climb trees to shake the fruit and take it to their burrows. No part of that sentence is even remotely true: hedgehogs can't climb trees, they don't eat fruit, and they do not hoard food. And yet the piece of zoological misinformation has likely been perpetuated, unchecked, since the times of Pliny the Elder, for 2,000 years.

Luckily, on the whole, we do rather better these days in terms of our zoological knowledge. It therefore seemed high time to produce a brand-new version of a bestiary, celebrating the extraordinary wonders of the animal kingdom, based on modern study not hearsay. Unlike its medieval counterparts, *The Modern Bestiary* is based on solid, peer-reviewed research—although you might find the occasional quote by Pliny and Aristotle within it—and the beautiful illustrations by Jenny Smith are accurate.

In addition, rather than using animals as a pretext for lessons on morality (I don't think I am qualified to moralize), they are used here as a pretext to showcase biological concepts. After all, if we tried the medieval approach of interpreting human nature quite literally through the prism of animal behavior—but leaning on today's extensive body of knowledge—we would soon run into difficulties. Ethics based on what is "natural" might open the floodgates for justifying cannibalism (coconut crabs do it, see page 26), incest (just look at bees, page 158), or sophisticated brainwashing (see those master manipulators, flukes and wasps, pages 98, 180 and 212). On the same note, it's interesting that people often try to prove a somewhat loaded or politicised point by referring to the animal kingdom—for example, that we are/aren't supposed to be vegetarian, or monogamous, or homosexual, or transsexual, or living with parents for however long. In this book, I'd like to show that pretty much any such point can be proven or disproven if you dig into the animal kingdom deep enough. Sacrificial mothers? Sure. Terrible mothers? Sure. Paternal care? No problem. Infanticide? Yep, it's there. Same-sex couples? You got it. Animals changing sex? Of course. The diversity of adaptations and strategies is astounding. Because of that, while medieval authors used the natural world to look at human nature, this book uses it to look at, well, nature.

And what a lot of it there is! Precisely because the animal kingdom is so varied, selecting a mere one hundred species for this book was incredibly difficult. The selection was a bit like picking kids for a very large football team—except there are about a million kids to choose from, you know some of them really well, some fairly well and some not at all; a few show up almost when the trials are over, only to turn out to be fantastic at football; on top of this, there are millions of others actively hiding from you.

At the time of writing, there are over 1.4 million described animal species indexed in the Catalogue of Life, an online database sourcing information from peer-reviewed, scientifically sound sources. This number, though impressive in absolute terms, is still rather modest compared to what we don't know: estimates for the *total* number of species on Earth range from 8 million to 163 million. Out of the catalogued species, the vast majority are arthropods (1.1 million species) and, within those, insects (over 950,000). The vertebrates comprise barely 5 percent of all described animals, and the most charismatic taxa—birds and mammals—a measly 0.7 and 0.4 percent, respectively.

I am bringing up these statistics because chances are that I have not included your favorite species in this book. I apologize; the odds were against me. I aimed to make a selection encompassing both taxonomic and conceptual breadth, and I tried—believe me!—not to be too mammal-centric. While at first it was a difficult task for someone considering herself a mammalogist, as I progressed in my writing, I realized just how many exciting species I had been missing out on throughout all my years of biological education and research. Speaking to my colleagues, I noticed I wasn't the only one—as scientists, we tend to focus on a select taxonomic group, geographic area or biological problem, but we rarely get the chance (or reason) to have a major overview of animals across the entire kingdom.

Consequently, the process of writing a book has been a joyous journey of discovery. In fact, I recommend a similar exercise to every budding zoologist: try to pick a list of one hundred species that you would like to show to the world—and make sure that you justify each one, and that they are spread across different branches of the evolutionary tree. I say this, since had I written *The Modern Bestiary* during my undergraduate or master's degree, I might not have chosen to research mammals for my doctorate. In the grand scheme of things, other taxonomic groups are so much more interesting! Mammals are like a conservative, suburban family, with Mom, Dad, two-point-five kids, and a Labrador. Among mammals, hardly anything is unusual: there are only two sexes, the body plans are generally similar, everyone uses lungs to breathe and has a pretty stable body temperature. By contrast, other taxa can do pretty wacky stuff—change sex, regrow limbs, produce amazing chemical substances, age in reverse . . . Think of the mole salamanders (page 48), who puzzle taxonomists by forming a female-only species. Or humble roundworms (page 74), who can force ants into impersonating fruit. There's a huge spider (page 72) that keeps frogs as pets. And a sea slug (page 136) that decapitates itself, leaving its head to live happily ever after on its own. How about butterflies (page 188) that are able to make crocodilians cry, only to drink their tears? All of these magnificent, outrageous—but under-studied and unknown—beings find their way into this book.

The trouble is, the public's attention is largely biased toward vertebrates—mainly mammals, birds, and a handful of lucky reptiles, amphibians, and fish. This is understandable, as it is generally easier to relate to, and empathize with, a chimpanzee than a sponge. In this book, I tried to touch upon different branches of the phylogenetic tree but of course I acknowledge that the animals placed on its more remote twigs are generally studied in less detail—either because of lack of scientific interest, or because of difficulty in accessing them, or perhaps due to the species' rarity—even though it is precisely that lack of relatedness to us that makes them so interesting. They have senses we cannot even comprehend, abilities we only see in sci-fi movies, and they live in places that have hardly been explored yet. And each of these unique weirdos has their own dramatic story, some of which I have attempted to tell here on their behalf.

Even so, the stories are not so much about the animals themselves—though, yes, there are a few chapters about species that are simply cool, no strings attached—but mostly they are an excuse to discuss ecology and evolution. Bear in mind that I have very consciously focused primarily on behavioral ecology (what animals do, how they behave, what are their relationships with other organisms), rather than going into conservation issues in great depth. Conservation is a study of people as much as, if not more than, animals, and this book aims to put the animals themselves front and center.

This is not to say that conservation is not important, of course; on the contrary, it is terribly important. By writing *The Modern Bestiary* I wanted to share my exuberance at the astounding diversity of wildlife. But being able to continue the glorious exploration of just the weirdest and most disturbing stuff that animals do depends on whether these creatures are still around—and whether they have a place to live. In a world with soaring human populations and larger encroachment into wildlife areas, human–wildlife conflict is inevitable. And, when it happens, more often than not, wildlife comes off worse. Because of the increasingly frequent interactions between humans and wildlife, there is a bigger potential for the spread of diseases (and don't we know what happens in that scenario!). The growing pressures to turn natural areas into croplands, cities, plantations, or mines lead to habitat loss—one of the main causes behind biodiversity decline. Changes in land use (along with burning fossil fuels) are also behind climate change, the other huge driver of biodiversity loss. Its consequences—rising temperatures, changing rainfall patterns, ocean acidification, extreme weather events—are too sudden for wildlife to be able to cope with. Animals living in colder areas or higher altitudes won't have anywhere to move when temperatures reach levels beyond their abilities to survive; the same holds true for marine species, which additionally are faced with changes in water pH. Weather is becoming too unpredictable to ensure fruitful breeding seasons, while higher frequencies of fires, droughts, and storms are already taking their toll on previously threatened wildlife. Pollution (chemical, light, sound) will further disrupt breeding cycles. As people travel and move goods around the globe, they also move animals, deliberately or otherwise, which has the potential of threatening native flora and fauna. And overharvesting of wildlife—be it for food, medicine, decoration, or pleasure—will lead to more population declines.

Even though this is a grim picture, there are some solutions in sight, though they require the attention of, and contributions from, governments around the world. In 2012, a report by environmental economist Donal P. McCarthy and colleagues, published in the journal *Science*, estimated that the costs of improving the conservation status of all the threatened bird species in the world would amount to $1.23 billion annually over a decade. The calculations were based on existing efforts, as well as previous, successful projects. Some costs accounted for single-species approaches, such as captive breeding; others included habitat preservation and restoration, which will likely benefit multiple species. While the total bill may sound like a lot, the entire sum would comfortably be covered by what Brits spent on sweets and ice-cream the year the report was published ($15.27 billion); the annual cost also compares favorably with money generated by the top-grossing movies each year. Still, ten years down the line, there has been no systemic, collective effort to throw money at the problem. Perhaps the current rise in environmentally focused grassroots movements will help to make it a higher priority for the world's governments.

In the meantime, for positive examples of bottom-up conservation projects, it is worth looking at Conservation Optimism—a global community highlighting positive change and fostering effective action. Its underpinning motivation is that hope and optimism can provide motivation for making a change; perhaps more effectively than the overpowering doom and gloom. I share that view, and so my hope is that this book will inspire you to discover the joys of the natural world. Perhaps it may lead you to explore how you can help nature in more practical ways. A good place to start may be joining a local wildlife group, engaging in citizen science projects—such as bird or bumblebee surveys—or learning how to make your surroundings more wildlife-friendly. These small undertakings allow you to connect with nature, while also helping conservation research. At the same time, try to be more conscious of your lifestyle—from where you travel and what you buy, down to what you post on social media (see the slow loris chapter, page 68, for an explanation of why it matters). The world we live in is astonishingly well connected, and your actions, choices, and decisions can make an impact on the other side of it.

• • •

But let's get back to the book. Anyone who is friends with one will confirm that zoologists can be counted on to lower the tone of every conversation. Right now, after a year of working on *The Modern Bestiary*, my search engine results also reflect this. From cuttlefish sex, to sea cucumber bums, to panda lactation, I spent hours upon hours researching the grossest and strangest aspects of animal lives. And I loved every minute of it.

Here is what I have learned throughout the writing process:

- Microsoft Word's thesaurus is too prudish to find alternatives for "having sex."
- Ditto "fart."
- Ditto "butt."
- Ditto "poop."
- However, it insists on correcting "monotremes" to "monodramas," leading the world to reconsider the artistic abilities of the platypus.
- Species' scientific names really come in handy, especially when one is attempting to look up papers on herring communication, and the key expert in computer communications is called S. C. Herring.
- The frustration of a person who spent two days writing a chapter about prairie dogs' sophisticated language only to find that the research has been debunked the year before can only rival the frustration of the authors of that research.
- Finding information on whether bees are incestuous proves tricky, because every search engine swaps "bee incest" for "bee insect,"
- Meanwhile, searching for images of "swift + toes" produces innumerable pictures of Taylor Swift in sandals.
- There is no shame in calling long-lost chiropterologist friends only to ask them about the details of bat genitalia.
- There is also no shame in calling long-lost classicist friends only to confirm that *dolichophallus* does indeed translate as "long penis."
- My fellow researchers are heart-warmingly helpful—no call for support, clarification, or resources was left unanswered. I received tips from a herpetologist in Canada and a palaeontologist in China, an entomologist in Poland and an evolutionary ecologist in Sweden; I was even able to call an ornithologist stranded on Isla Róbinson Crusoe on his way to do fieldwork. I (twice!) spent a friend's party discussing moths and bats with a new acquaintance who turned out to be a biodiversity specialist. And I was flooded with what looked like inappropriate photos until I realized that they actually depict priapulid worms.

As I imagine this list makes clear, zoology as a discipline is enormous fun. I mentioned earlier some of the difficulties of choosing which animals should be included in this book. In the end, I hit upon a selection method based on that of Marie Kondo. I would look at each species and ask myself: does the idea of writing about this creature spark joy? I hope that you experience the same joy learning about my chosen animals as I did writing about them.

The Modern Bestiary was conceived in 2021, an otherwise difficult, pandemic year. Focusing on the manuscript provided a much-needed portion of escapism, as I voraciously gobbled up animal facts, stories, and delicious details. That year, another creative process took place: halfway through writing my book, I produced a small mammal of my own.

As I was nourishing both the flesh and the paper baby, my older daughter, who is five at the time of publishing, was trying her hardest to help me with both. She assisted with bath times and changing nappies on one front, and, on the other, insisted on ghost-writing chapters for me ("A PLATPS IS NOT A BARD BAT IT STIL LEYZ EGZ") and "improving" the illustrations by diligently coloring them in. And while I did not take many species requests, the Amazon river dolphin (see page 86) was her idea.

I hope you enjoy *The Modern Bestiary*—if you do, with the number of currently described species, there is potential for 14,000 sequels.

NOTE

Humans in general—and scientists in particular—love making sure that everything is neat, orderly, and categorized in little boxes. In biology, the process of putting animals in little boxes, concept-wise, is called taxonomy. The taxonomic hierarchy, in biological classification, involves ranks: species, genus, family, order, class, phylum, kingdom, domain. While this all sounds nice and neat, very often animals don't let themselves be taxonomically pigeon-holed. In this book I therefore use a broader taxon at times—be it because I have a hard time choosing just one species from a genus, family, order, or phylum, or, as in the case of the mole salamander, a species is not really an adequate concept, since their classification is a bit more fluid.

Where I mention more than one species, note that "*Antechinus* spp." means "several species in the genus *Antechinus*." By contrast, "*Antechinus* sp." would indicate a single, unspecified species of the genus *Antechinus*.

The species in this book are arranged alphabetically in three sections, based on elements the wildlife can be found in: Earth, Water, and Air. I appreciate that many animals straddle more than one of these habitats, and sometimes may even feature in all three, but through this arrangement I wanted to demonstrate the diversity of taxa that share some physiological, behavioral, or ecological adaptations.

If you are interested in how the species are related to each other, please refer to the inside of the cover for a sketch of the tree of life with major taxonomic groups highlighted.

EARTH

Antechinuses

Antechinus spp.

You would be forgiven for confusing the small, long-tailed, pointy-nosed and beady-eyed antechinus for a mouse—yet this little critter is more closely related to the koala than a rodent. Antechinuses, also known as pouched mice, are Australian marsupials; they range in size between 16 and 120 grams and belong to one of fifteen species in the genus *Antechinus*. Their diets consist of invertebrates and occasionally fruit. When food is scarce, pouched mice resort to hunting small birds or mammals; during feeding they turn their victim's carcass inside out and leave a neatly everted skin as a testimony to their feast. As if that was not gruesome enough, male antechinuses also have the most macabre love lives. They, quite literally, mate themselves to death.

In the winter, pouched mice embark on a synchronous—and very vigorous—mating extravaganza, during which the males focus solely on copulation, dismissing such trivialities as eating or resting. They have one goal: to pass on their genes. Their devotion to this goal is ultimate. Male antechinuses might not be particularly romantic (their courtship is practically nonexistent: the object of their desire is grabbed by the nape of the neck and subdued so forcefully that mated females might have patches of hair missing from their backs), but when it comes to endurance, they are like Duracell bunnies. A single act of antechinus copulation can last up to fourteen hours—and that's just the first mating . . .

After one to three weeks of such exhausting orgies, the male bodies start to give in. Constant mating and fighting off other determined and lustful competitors puts a lot of pressure on the males, leading to a testosterone-fueled increase in stress hormones, corticosteroids. Uncontrollable floods of corticosteroids weaken the immune system, leaving the pouched mouse playboys susceptible to parasites in the blood and intestines, bacterial infections of the liver, as well as gastrointestinal ulcers and internal bleeding—eventually leading to death. Within a couple of weeks, due to the massive male die-offs, the antechinus population is reduced by half.

The do-it-and-die reproductive strategy, known as semelparity (from the Latin *semel*, "once, a single time," and *pario*, "to beget"), makes evolutionary sense. The massive breeding sprees happen all at once so that the females can raise their babies at a time of maximum food availability in spring or summer—and the males are not competing for the nosh with the younger generation. What gets them in the synchronized mood for love is the sun—or, rather, changes in daylight hours: lengthening days indicate that, in a few weeks, food will be plentiful. Since peak food availability varies in different antechinus ranges, the mating seasons also differ between species. Interestingly, if several species inhabit the same area, their breeding seasons are staggered to avoid direct competition; the bigger species, such as *Antechinus swainsonii*, breed earlier than the smaller ones, like *A. agilis*.

While the males place a special emphasis on the "till death do us part" aspect of their relationships, the females may live long enough to raise a second, or, very occasionally, third litter—each time as single mothers. The 6–13 young are born tiny and helpless, and since the antechinus pouch is only a rudimentary flap of skin, the offspring dangle like grapes attached to their mother's teats. For about five weeks she will drag them around awkwardly, as they hold on for dear life. A month and a half later, the youngsters become independent—and the following winter the morbid dating game starts once more.

Aye-aye

Daubentonia madagascariensis

Knock, knock! Who's there? It is aye! I who? Aye-aye, a creature so abhorred and feared that its real name is never spoken.

In fact, the common name of the *Daubentonia madagascariensis* is shrouded in mystery. Unlike the names of other lemurs, it is not onomatopoeic—it doesn't mimic the noise produced by the animal. One hypothesis states that the word comes from the Malagasy *heh heh*, meaning "I don't know"—likely said to avoid speaking the name of this ominous primate. Aye-ayes are believed to be bad omens—either harbingers or causes of death—and just being looked at by one can lead to trouble.

To be fair, seeing an aye-aye (especially at night, when they are active) could give anyone the heebie-jeebies. In contrast to other lemurs, which tend to be fluffy, cute and generally easy on the eye, this creature has the air of a cat struck by lightning. French colonists of Madagascar thought it was made by the devil, since "it has the teeth of a rabbit, the hair of a pig, the tail of a fox, and the ears of the bat." They raised valid points: the ever-growing, rodent-like teeth led zoologists to initially believe that they were seeing a large squirrel. Its fur is shaggy, unkempt, and graying-black, with the hairs of the long, bushy tail reaching over 20cm. The ears are oversized. And then there's the evil eye—or rather two: the wide-set, ogling headlights of a nocturnal mammal give the aye-aye a somewhat deranged look.

But perhaps the eeriest feature is the hands. This lemur's hands are proportionally the longest of any primate, accounting for 41 percent of the forelimb. The aye-aye's fourth finger is particularly lengthy—over two-thirds of the total length of the hand—although the third finger is the most unusual. As skinny as a wire, it is incredibly maneuverable thanks to a ball-and-socket joint; it moves independently of the other digits, and can be bent at a right angle to the back of the hand. It is so thin that the animal normally rests it against the fourth digit, thus keeping its fingers permanently crossed.

The special middle finger is used for knocking on wood—a behavior known as percussive foraging. In total darkness, the aye-aye taps the nail of its creepy middle digit on branches and tree trunks—up to an impressive eight times per second—and uses its bat-like ears to listen for hollow cavities containing grubs. Then, with its sharp teeth, it gnaws through the wood surface and uses the all-purpose middle digit to extract larvae. In doing so, the aye-aye fulfils the ecological role of a woodpecker, a bird absent in Madagascar.

Aye-ayes don't just feed on grubs—they are omnivores, and will happily eat fruit, nectar, nuts, and fungi. Their varied diet explains one more interesting adaptation: they can hold their liquor. When given the choice of nectar-like drinks with varying alcohol content, these lemurs choose the booziest ones, likely because they are used to eating the slightly fermented nectar of the traveler's palm, which they yet again extract with their poke-y finger.

In villages in some parts of Madagascar, it is thought that the aye-aye may slit a person's throat with the thin finger—or that when it points at a person, it foretells their death. The animal is believed to be so unlucky that it is sometimes killed on sight, and even hung upside down by the roadside, to ensure that passers-by take the bad luck with them away from the village. Unfortunately, with rapid habitat loss, aye-ayes find themselves closer and closer to human settlements, especially since they are tempted by the mangoes, sugarcane, or avocados growing on village farm plots.

Of course, crop raiding does not help the public image of the already loathed and dreaded primate. Yet, far from being a supernatural harbinger of doom, it seems the aye-aye is merely a disheveled, crazy-eyed drunk, wandering about looking for finger food: not a creature you'd want to meet on a dark night, perhaps, but completely harmless.

Banana slugs

Ariolimax spp.

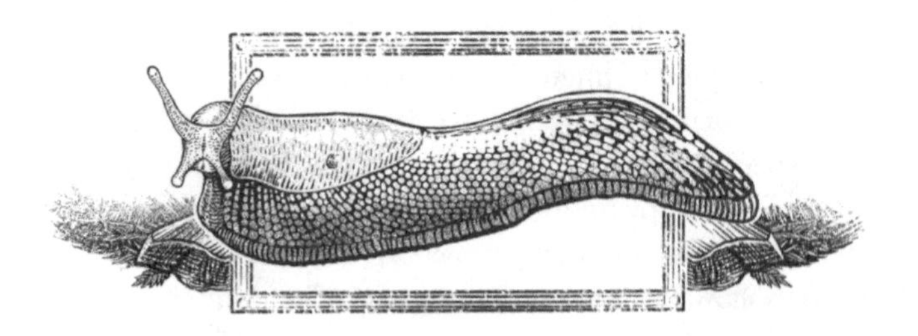

What measures 6–10 inches, looks like a banana, and is very well lubricated? That's right—it's the banana slug! Did your juvenile mind suggest other answers? Don't worry, it simply means that you think like a zoologist: a scientist who likes few things more than looking at weird animals' weird genitalia.

There are five species of banana slugs (genus *Ariolimax*); all inhabit western North America, favoring humid coastal forests from Alaska to California. As decomposers, they form an important part of the redwood ecosystem, feeding on fungi, detritus, dead vegetation, feces, and carrion. The slugs come in all shades of banana ripeness, from greenish or bright yellow to spotty, brown, or almost black. Like bananas, they change color with age—but also with moisture, food, and light. Yet unlike bananas, banana slug species are mainly differentiated by the structure of their private parts.

As with most other gastropods, these fruit-like molluscs are simultaneous hermaphrodites: each slug is the proud owner of a muscular vagina and oviduct, as well as a hollow, retractable penis. In some instances, the organ can be as long as the slug itself, giving rise to the scientific name of the slender banana slug, *A. dolichophallus*—literally translated as "long penis." Still, nineteenth-century American slug researchers Henry Pilsbry and Edward Vanatta reported the discovery of a strange new species—one that lacks a phallus altogether; they called it, rather unimaginatively, *Aphallarion*. Little did they know that *Aphallarion* is not a separate species, but a regular ol' banana slug who just had a bit of a mishap during lovemaking—something that happens to about 5 percent of the *Ariolimax*.

In general, unlike the aggressive copulations of bed bugs or flatworms (see pages 28 and 132), banana slug sex is a peaceful and cooperative affair. Two slugs align themselves in a 69 position, wait for their genitals to swell and spring from the pores on their heads (let's not stoop to dickhead jokes here, please), and then use either one or both penises to connect and exchange sperm. After the act, which may last for hours, the penises are either released smoothly, pulled out after a lot of stretching, yanking and turning, or . . . they get stuck for good. Faced with such an awkward situation, banana slugs resort to the ultimate solution: apophallation, or chewing off the partner's phallus.

Since slugs are generally slow-acting creatures, apophallation takes several hours. The procedure causes much pain, judging by the recipient's frantic writhing—predictably, as the penis is sluggishly scraped off with the castrator's radula (a ribbon-like, toothed tongue), and starts withering away throughout the lengthy process. The animal who severed the stub ends up eating it; clearly there is no use in letting all that pain go to waste. The apophallated individual is unable to regrow the penis, but it can still breed as a female, investing energy in egg production. This could be advantageous to the castrator, as increasing the density of female-only slugs may reduce competition. Sometimes, if both slugs are stuck (or perhaps vengeful?), apophallation is reciprocal.

A key component of the slugs' courtship is lubrication, as they use slime trail pheromones to attract mates. Like KY jelly, banana slug lube comes in packets: small granules containing mucins, proteins which expand their volume a hundredfold upon contact with water. It is simultaneously a lubricant and an adhesive, and has both solid and liquid properties. The mucus is incredibly useful, not only in finding love, but also in locomotion, defense against predators (a mouthful of slime tastes gross, and it numbs the tongue) and even surviving periods of drought.

Whether it was the exciting physics of slug slime, or their propensity for love bites, banana slugs have been chosen as the official mascot of the University of California Santa Cruz—make of that what you will.

Bat-eared fox

Otocyon megalotis

Take some shaggy fur, four long legs, a bushy tail, the face of a graying miniature pinscher and a pair of oversized ears—and you will have a bat-eared fox, one of the most unusual members in the canid (dog) family.

The first things that stand out when you spot this fox are the ears—they measure 11–13cm, and can make up a third of the animal's height. They are so prominent that they feature in the species' scientific name twice. The generic name *Otocyon* comes from the Greek *otus*, "ear," and *cyon*, "dog," and the specific name, *megalotis*, is again derived from Greek: *mega*, "large," and *otus*, "ear." Eared dog with big ears. But why are these lugholes so big?

One reason is thermoregulation. The ears, with a relatively thin skin and numerous blood vessels near the surface, help dissipate the heat. The other reason (best said in the voice of the Big Bad Wolf from "Little Red Riding Hood") is that they are "all the better to hear you with." Bat-eared foxes feed on termites, and rather than sniffing them out or looking for their prey, they listen out for them. Even though termites are mute, the large, gray bat-ears are sensitive enough to pick up on their movement underground.

The Dumbo dogs' crunchy, insect-rich diet requires a lot of chewing. Thankfully, bat-eared foxes have a mouth jam-packed with pointed teeth—more so than most other mammals: up to fifty. They also have a specially modified digastric muscle in the jaw that allows them to open and close their mouth extraordinarily quickly, up to five times per second. When not munching on insects, they forage on fruit (especially berries), vertebrates—such as lizards, small mammals, and nestlings—and they will scavenge on a carcass if they come across one.

There are two subspecies of bat-eared foxes; one found in eastern Africa, the other in the south of the continent. The former subspecies is mainly active at night—again, relying on big ears is helpful if you are trying to hunt in the dark—but the latter becomes diurnal during winter, likely because termites are more active during the warmer parts of the day. Interestingly, when feeding during the day, the foxes attract the attention of numerous insect-eating birds, such as ant-eating chats (*Myrmecocichla formicivora*), crowned lapwings (*Vanellus coronatus*) or northern black korhaans (*Afrotis afraoides*). Birds pick up on the fact that the foxes detect insects, and treat their presence like a restaurant noticeboard. Thankfully, squabbles are rare—there is enough food for everyone.

Bat-eared foxes are social animals; they live in groups of two to fifteen, usually consisting of a monogamous pair and grown-up young, or stable family groups of a male and up to three females with pups. The group sleeps and forages together, and members frequently groom and play with one another. The dads are very hands-on when it comes to looking after the young: they are involved in all aspects of childcare except nursing. For the first fourteen weeks after birth, the fathers are the ones devoting significantly more time than the mothers to guarding the den and pups. Bat-eared Papa will huddle with the babies, clean them, carry them and chaperone them during foraging trips—in contrast, Mama will spend most of her time away from the den, feeding, to maintain milk production. These friendly, gentle canids are big on teamwork as well as being large in the ear department.

Brown rat

Rattus norvegicus

Despite their Latin name, brown rats (*Rattus norvegicus*—Norway rats) are originally from Asia, most likely China or Mongolia. Thanks to their adaptability, they have spread to all continents except Antarctica, and are present almost everywhere humans are. Our relationship with these rodents is a complex one: they are pests and companions, carriers of disease and model animals for laboratory research; staple dishes and food taboos—and because of their ubiquity, they play a role in the myths and stories of peoples all around the world.

In Chinese astrology, the rat is the first of all zodiac animals. According to legend, the order was decided by the sequence in which the animals arrived at the Jade Emperor's party. The rat tricked the ox into giving him a ride, but just as they reached the Jade Emperor, the rat jumped off, landing ahead of the ox and winning the race. The rodent thus became a symbol of intelligence, ambition, and ruthlessness. However, while brown rats certainly are very intelligent, it turns out that they are much more empathetic than the Chinese story would have us believe.

How might we measure the empathy of rats? A research team from the University of Chicago tasked the rodents with liberating fellow rats trapped in small restraining cages. The laboratory rats repeatedly freed their ensnared friends, even when there was no particular reward associated with the task. Indeed, when presented with a choice: to free a cage-mate, or to open a container with chocolate chips (a beloved ratty snack), rats chose to do the former. Not only that, they then shared the chocolate bounty.

Similar behavior was observed when rats were in a more stressful situation: being soaked in water. Rats find soaking very distressing, and in the experimental set-up they could only escape from a pool if a fellow rat opened a door leading to dry land. Once again, when rodents were faced with the choice of liberating a companion or accessing chocolate cereal, they chose to help. What's more, if the savior-rat had previous soaking experience, it would release the cage-mate much quicker. Empathy galore!

But does the empathy only extend to rats? Would a rat save a non-rodent creature from entrapment? While this has not been tried specifically with oxen (to disprove the Chinese myth), a study from the University of California San Diego examined whether rats could make friends with . . . robots. The robots were small, mobile, rat-shaped devices, and were either "social" (remote-controlled to interact with rats, play with them or release them from restraining cages) or "asocial" (moving in a pre-programmed, randomised manner). As with real rats, the experimental rats "freed" the robots from traps—and were much more likely to do so for the social ones.

Rat interactions with technology are not limited to frolicking with robots. Researchers from the University of Richmond in Virginia taught rats to drive miniature cars. The ratmobiles were not overly complex—they had an aluminum floor and three copper bars for steering. When the rat chauffeur grabbed the copper bars, it completed an electrical circuit that started the car's motor. Touching the left or right bar allowed the driver to change direction, while the middle bar propelled the vehicle forward. With the help of treats (mini marshmallows), rats learned to drive over progressively longer distances, turn the vehicle and park near a food source. What's more, rats were not just skilled drivers—spending time behind the wheel actually relaxed them. The research team measured hormone levels in the animals' droppings—in particular, the stress hormone corticosterone, and dehydroepiandrosterone, which counteracts stress. The longer the rats took driving lessons, the higher the proportion of dehydroepiandrosterone in their feces, indicating higher resilience to stress. Interestingly, rats who were passengers of remote-controlled cars turned out to be more stressed than rat drivers. Clearly, anxious backseat drivers are not just a common phenomenon in our species.

Caecilians

order Gymnophiona

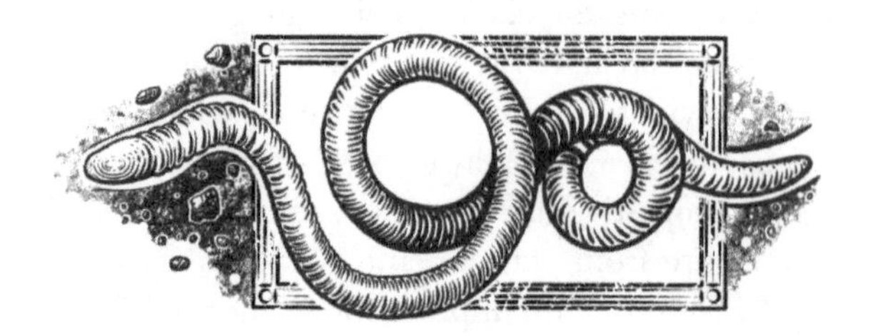

Among amphibians, frogs, and toads are the ones who receive the most publicity. Salamanders and newts get the occasional mention. But hardly anyone knows that there is a third order of amphibians: the caecilians.

What are caecilians? Imagine a super-sized earthworm, but with a backbone. While the smallest of these creatures, about 10cm long, can be mistaken for worms, the largest ones, such as Thompson's caecilian (*Caecilia thompsoni*), reach a solid 1.5m in length and resemble snakes. Unlike snakes, they are not covered in scales—their skin is smooth and slimy. But caecilians do share a specific trait to accommodate for their long, thin body shape: as with snakes, their right lung is bigger than the left. The sole exception is the single lungless species, *Atretochoana eiselti* (due to its looks also known as "penis snake," though it is neither penis, nor snake), which breathes directly through its skin.

Caecilians are found in the tropics, in South and Central America, Africa, and southern Asia; they feed on invertebrates, such as earthworms, termites, and ants. Because they live underground or underwater, they do not have great eyesight, and can probably only just determine light from dark; in some species, the eyes are rudimentary. This is reflected in their name, which stems from the Latin *caecus*—"blind."

Perhaps surprisingly, these unseeing, obscure critters can make very devoted mothers. While most caecilians are viviparous—that is, they give birth to live young—about a quarter are oviparous, or egg-laying. Female chikilids, recently discovered Indian caecilians in the family Chikilidae, guard their developing eggs for two to three months, apparently not eating at all during this period. Other species take maternal care a step further, by offering their developing babies a nourishing, if somewhat unappetizing, dining option: their own skin.

The skin of brooding females doubles in thickness to provide a nutritious meal that, like mammalian milk, is rich in fats. After hatching, baby caecilians are treated to a skin banquet; it may be their sole sustenance for the first few weeks of their lives. Young caecilians use specialized, spoon-like baby teeth to rip off shreds of skin; the mother remains calm during that time, as her offspring writhe and squabble over the most delicious morsels. The feast occurs every few days, giving maternal skin time to regenerate. As the little ones grow, their mother loses weight (some 15 percent in a week).

Skin-feeding behavior, known as dermatotrophy (from the Greek *derma*, "skin," and *trophy*, "nutrition"), was first observed in the Taita African caecilian (*Boulengerula taitana*) from Kenya. On the other side of the world, in South America, two other caecilian species also offer their hides to their young. Additionally, South American ringed caecilians (*Siphonops annulatus*) top up this nourishing meal with an even more tempting supplement: two types of excretions from the mother's cloaca (backside), one watery, the other more viscous. This fact is perhaps worth mentioning at dinner time, if your children are ever fussy about eating their Brussels sprouts.

The lifestyles, habitats, and biology of caecilians are woefully under-studied. According to the Red List of Threatened Species, published by the International Union for Conservation of Nature in 2020, a whopping 40 percent of amphibian species are threatened. At the same time, around 17 percent are classed as "Data Deficient;" in other words, lacking enough data to assess their extinction threat. Unfortunately, this does not mean that they are safe from threats; merely that there is not enough interest in, and knowledge of, these animals to evaluate how the species is doing. For caecilians, this lack of information is even more pronounced: out of the 183 species present on the IUCN Red List, 52 percent (95 species) are Data Deficient. It would be a terrible shame to let these fascinating animals go extinct before we find out more about them.

Coconut crab

Birgus latro

Gather round for the story of a fearless buccaneer, causing terror in a tropical paradise. But be warned: it's not a tale for the lily-livered.

It is, in fact, a story of islands, palm trees, and a ruthless crab, huger than any other. Of a crab that, in true pirate fashion, intimidates, plunders, and kills. Behold: the coconut crab, *Birgus latro*. Weighing 4 kilos, and with a leg span of over a meter, this goliath crustacean is both the largest terrestrial arthropod and the largest invertebrate living on land. Western scientists first heard of the species from privateer Francis Drake, who became acquainted with it on his trip around the world. Meanwhile, Charles Darwin described the crab as "monstrous"—and it certainly is menacing. These strongmen of the invertebrate world can lift as much as 28kg, and the powerful claws of the largest individuals have an estimated pinching force of 3,300 newtons, which is comparable to the bite of a hyena.

Coconut crabs are related to hermit crabs—and, true to their ancestry, juveniles reside in gastropod shells. However, once fully grown, they leave their shell homes and rely on their hard, calcified exoskeletons and massive pincers for protection. Indeed, it is the coming out of the hermit shells that accounts for the coconut crabs' huge claws: leaving behind the restrictive snail-bunker allows their bodies (pincers and all) to reach impressive sizes—and strength.

The mighty claws are used not only for self-defense, but also for obtaining food (which is located using the sense of smell). Coconut crabs, as their name implies, are able to open coconuts—rather than walking the plank, they walk the palm tree, gripping on to the trunks with their limbs as they climb. While these crustaceans usually feed on fruit, nuts, seeds, and other plant material, they are omnivores and will happily eat meat given the chance. If no scavenging opportunity presents itself, they hunt. The crabs have been reported to kill rats, as well as attack birds in their sleep, breaking their wings with their deadly grip and eating them. With a forceful handshake, the crabs are able to easily crush even the largest bones. And they are not opposed to a bit of cannibalism, either.

When not killing defenseless birds, the giant crustaceans pillage. Another name for the species is the robber crab, or palm thief—unsurprisingly, as they frequently snatch items from humans and carry them off. They have been known to steal pots, shoes, watches, and cameras; there are stories of crabs pinching bottles of whiskey, and, in a truly swashbuckling power move, one allegedly even seized a gun from a military guard. It is unclear why they steal so much—it could be because novel items have interesting smells and need to be investigated as potential food.

Robber crabs inhabit coral atolls across the Indian Ocean and parts of the Pacific—including Christmas Island, where they are present in the highest density. They probably reached all the remote islands as larvae, since only then are they sea-bound. After three to four weeks at sea, young crabs move to land; once grown, they lose their ability to swim. Breeding females need to be careful when releasing their eggs into the ocean at high tide, since if a wave knocks them into the water, they drown and become shark bait.

On one of the islands they have conquered, the palm thieves might have earnt another claim to notoriety. Some historians hypothesise that coconut crabs could have devoured the famous aviatrix Amelia Earhart after she crashed somewhere over the Pacific Ocean. The evidence is not conclusive—yet with the robber crabs' pirate attitude, nobody should put man-eating past them.

Common bed bug

Cimex lectularius

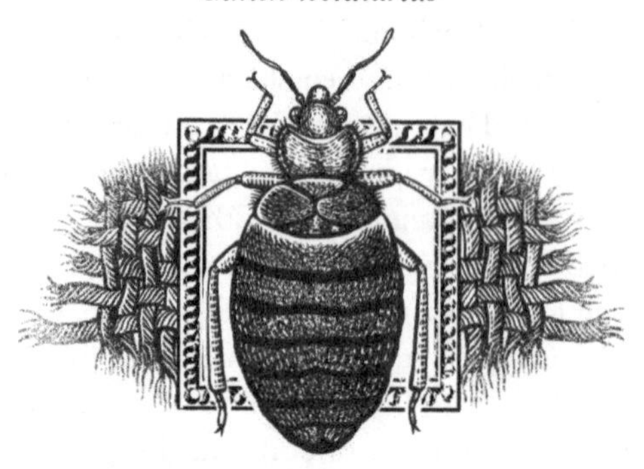

What is the surest way to annoy an entomologist? Calling something a "bug" if it is not, in fact, an insect belonging to the order Hemiptera (i.e. true bugs). The bugs' tell-tale sign is a very distinctive "beak," or rostrum—a mouthpart modification that enables piercing and sucking. While most true bugs use this apparatus to extract plant sap, there are some who prefer a less vegetarian diet: the blood of birds and mammals.

Unlike other "bugs" who might not be bugs (for instance, lovebugs are really flies, and maybugs are beetles), common bed bugs, *Cimex lectularius*, are named appropriately: they do belong to Hemiptera, and they can often be found in beds in hotels, hospitals, or trains around the world. They have been pestering humans for over 4,000 years, and were mentioned in the works of Aristotle and Pliny.

These small, brown, flat insects use their buggy beaks to pierce the skin of their hosts. To make the feeding as efficient as possible, their saliva contains anticoagulants and vasodilators which speed up the blood flow. They also inject painkillers, to be kind . . . or more likely to avoid getting swatted.

Bed bugs search for hosts based on three main cues: temperature (warm-blooded animals are the easiest to detect), carbon dioxide concentration (from the air breathed out by the unsuspecting victims), and various body odors, such as sweat or sebum. Unlike many other invertebrate parasites, bed bugs don't live on the host, but only make brief contact during feeding and then scuttle away to a safe hiding spot, where they can avoid getting accidentally squashed. Meals take place roughly once a week, but are rather substantial, since a bed bug can triple its weight after a blood buffet. In extreme circumstances, the insects can survive for around five months without food. Interestingly, their gruesome tastes are valued by detectives: since human DNA can be recovered from a bed bug blood meal for up to sixty days, they are useful in forensics.

But feasting on blood, infesting human dwellings and helping identify dead people aren't even the creepiest things about this species. It turns out that the bed bugs' love lives could have inspired the works of the Marquis de Sade.

How so? Female bed bugs have a fully functioning reproductive tract—but male bed bugs don't make use of it during fornication. Instead, they employ their very sharp sex organ (suggestively named the hypodermic penis) to stab straight through the female's abdomen into her ovaries. This process can cause serious wounds and even lead to death, and is, very aptly, termed "traumatic insemination." Females have developed a special structure, called the spermalege, which acts as a shield to reduce wounds and infection. Sperm travels through haemolymph (the insect equivalent of blood) to seminal conceptacles, i.e. specialized sperm storage units, where it is squirrelled away until fertilization. Females lay several eggs a day, and hundreds in their lifetime, which means that a single mom could easily infest a previously bug-less area.

Courtship? Forget it! Bed bugs pick their mates on the basis of size: females are generally the larger sex, so for a male bed bug, big is beautiful. This means that when males detect anything that remotely fits the description (bed bug-sized and moving), they attempt to mount it. If a well-fed male happens to come along, he might fall victim to the passions of another male. Unfortunately, these attempts at same-sex coupling usually prove fatal to the recipient, probably due to gut injuries, as males lack the shield-like spermalege. Mounted males may trigger a "rape alarm," releasing the *Cimex* distress pheromone to repel sexual predators.

What about other forms of bed-buggery? Copulating with a different species? Sure thing, even though the chances of producing offspring are minuscule. Incest? No problem: bed bugs seem very resilient to inbreeding, and colonies often appear to have been started by a single mated female. When it comes to bedmates, bed bugs certainly don't bugger about! After all, this is probably one of the keys to their successful spread.

Common sexton beetle

Nicrophorus vespilloides

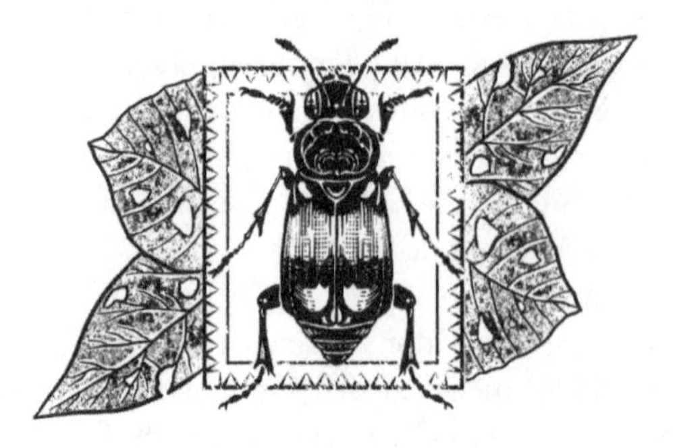

There are around two hundred species of burying beetles around the world, residing mainly in temperate regions. While formally they belong to the Silphidae family, burying beetles might as well have been dubbed the "arthropod Addams Family:" they are creepy, spooky, and altogether ooky.

Take one of the best-studied species of the group, the common sexton beetle (*Nicrophorus vespilloides*), a 1–2cm-long, handsome black-and-orange insect. Like a true Addams Family member, it is attracted to morbidity. Burying beetles are nature's undertakers; they can be found around dead animals, as they are necrophagous—they feed on carrion. The common sexton beetles locate the cadavers of small animals (usually rodents or birds) by using chemoreceptors in their clubbed antennae; they can detect a corpse within one day of death, several kilometers away. Because of their attraction to dead bodies, burying beetles play an important role in forensic research (like bed bugs, see page 28), as it is possible to estimate a time of death on the basis of the developmental stages of the insects found on the remains of the deceased.

Like the Addamses, sexton beetles form close-knit families. Mom and Dad work together to look after the young, which is an unusual occurrence among insects, who usually dodge biparental care. Actually, both females and males can also make very competent single parents, and males whose partner disappears can raise a brood as well as a single female or a pair. But whether single or coupled up, sexton beetles take an unusual approach to parenting: they use animal corpses as both larder and nursery.

Upon finding a carcass, the parents assess its size and state of decomposition. If the corpse meets their quality criteria, they will defend it against competitors—beetles as well as other carrion feeders. The pair will process the cadaver in preparation for their offspring—they bury it by digging beneath it and, as they do so, they snip off the fur or feathers. They shape the carcass into a ball, and once it is buried, the female lays her eggs around it. To inhibit decay—and the giveaway smell that could attract competition—sexton beetles impregnate the body-ball using oral and anal secretions that have antimicrobial properties. The burial prevents infestation by fly larvae, and the chemical conservation protects the carcass from fungal and bacterial decomposition.

When the larvae hatch, the mom and dad look after them, defending them (and the nutritious carcass) from predators. They chew an opening in the pre-prepared carrion ball to make it more accessible to the larvae, who feed directly from the gruesome pantry, and are also fed by the parents. The adults helpfully pre-digest bits of the carcass and regurgitate them to nurture the little ones. The larvae beg for food by touching the parents' mouthparts with their legs, and the younger ones do so more often than the older ones. However, begging comes at a cost—too much pestering may result in the offspring being eaten by the parent! Adults may adjust brood numbers based on the size of their stored carcass, to optimize the survival of as many offspring as possible—and the older larvae are more likely to survive, so they enjoy parental favor. Thanks to this strategy, the young are more likely to be honest in their signaling, and beg only when genuinely hungry. As a way of keeping the kids on their best behavior, it's unarguably efficient.

Common side-blotched lizard

Uta stansburiana

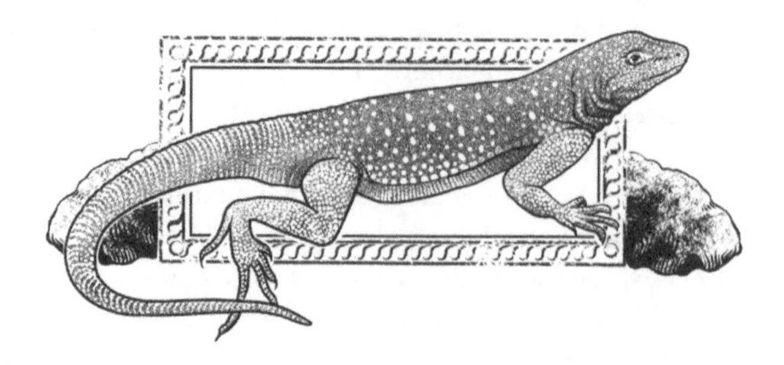

Even though love might be a game, few would imagine it as a game of rock-paper-scissors. And yet, the love lives of common side-blotched lizards (*Uta stansburiana*), small reptiles from the dry regions of western USA and northern Mexico, resemble just that.

Rock-paper-scissors is interesting because, despite there always being a winner within a pair (rock breaks scissors, scissors cut paper and paper wraps rock), there is no single overall winner—each option has strengths and weaknesses. Similar rules apply in the common side-blotched lizards' society. The males of the species come in three varieties, or morphs, easily differentiated by the color of their throats. The orange-throated males are the ultra-dominant ones: they are jam-packed with testosterone, aggressive, have big territories and secure multiple mates. The blue-throated males are mate-guarders: they are less aggressive, defend smaller areas and are protective of the females—but are vulnerable to cuckoldry by the macho orange males who usurp their territories. Finally, the males with yellow stripes on their throats don't bother with territories—instead, they resort to, as the evolutionary biologist John Maynard Smith termed it, the "sneaky fucker strategy.' As mimics of the yellow-throated females, they aim to slip unnoticed into the stomping grounds of the other two colored morphs for a spot of crafty copulation. Orange males are especially prone to being cuckolded, as their big territories don't allow them to be protective of all females at all times. However, blue males, who fiercely defend their mates, spot a sneaker straightaway and deter him. As a result, aggressive orange males steal the mates of the blue males, blue mate-guarders protect the females from the yellow males, and the yellow sneakers cuckold the orange machos.

Common side-blotched lizards rarely live past one mating season, so the winners of the evolutionary rock-paper-scissors are the ones who sire the highest number of viable, reproducing offspring. Yet the strange orange-blue-yellow stalemate leads to equilibrium: a different color morph dominates in the population over a five-year cycle.

Even though there is no single winner, male lizards have more angles to their game. Yellow-throated sneakers are sneaky on all levels: they exploit the fact that females store sperm, and outcompete the other two color morphs actually inside their mates. This is presumably because yellow males' sperm is long-lived enough to linger even after its donor has died, resulting in a higher number of posthumous offspring than the other two male types. On the other hand, blue-throated mate-guarders team up ("Us blue boys gotta stick together!") to defend their territories from the sneaky yellows. Interestingly, blue teammates are not relatives, even though they share genetic similarities. Blues who are neighbors are three times more successful at siring and producing mature progeny than the lonely blues. At the same time, the orange ultra-dominators use the opposite strategy: they choose sites as far away from the other oranges as possible to increase siring success (and decrease rivalry).

Meanwhile, female common side-blotched lizards, who can produce multiple clutches during a single season, play their own game. They come in one of two throat colors: orange or yellow. Orange females produce many smaller eggs, and the yellow ones generate fewer, bigger ones. The former are favored at lower population densities, because they produce more offspring, while the latter win at higher densities, as their young are of higher quality and therefore have the competitive edge. The females' game is played out in two-year cycles.

Still, not all populations of the common side-blotched lizards have all three color morphs. Genetic reconstruction shows that, although the rock-paper-scissors scenario has been present for millions of years, it has also been independently lost eight times—and doing so gave rise to new species or subspecies. The yellow morph is the one most likely to be lost—while there are some all-blue or all-orange populations, sneaker genes are not doing as well. It seems cheating at games doesn't always pay off.

European rabbit

Oryctolagus cuniculus

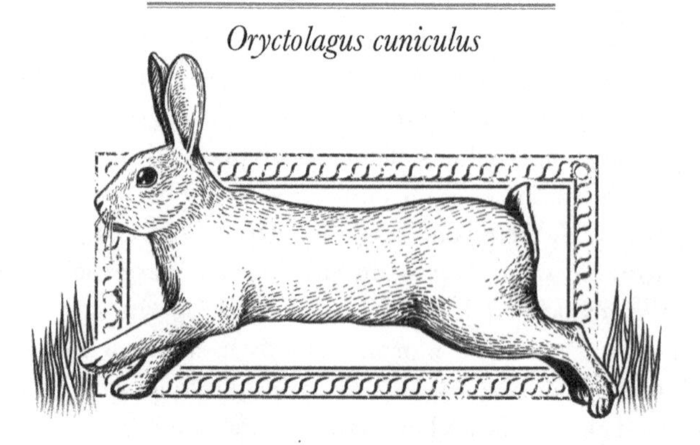

Even though they have large, permanently growing incisors, rabbits are not rodents. Since the early twentieth century, they have been classified in the lagomorph order, along with hares and the hamster-looking pikas. Although their front teeth are impressive, they differ from those of rodents—rather than two top incisors, rabbits have four, plus two stubby little teeth ("second maxillary incisors") right behind. This toothiness is a common feature of both rabbits and hares; but the two groups differ in their reproductive and survival strategies.

There are around thirty species of rabbits, and thirty-two of hares. Ecologically, hares adopt the "need for speed" approach to life. When in danger, they use their long limbs to outpace the chaser, reaching top speeds of 72 km/h. They are constantly on the run, as they live in open terrains, including deserts, grasslands, or tundra, with rudimentary shelter. Therefore, baby hares—leverets—are born precocial, i.e. quite well developed: covered in hair, with open eyes, and capable of movement on their own. They have to look sharp from the word go—their nursery is a mere depression in the ground, known as a form; their survival strategy is to make themselves invisible, and then bolt if busted.

By comparison, baby rabbits, or kittens, are born altricial: hairless, blind, and pretty useless. They can afford to be, as they are protected by an underground warren or a nest in dense cover. Rabbits, with their shorter legs and slower-paced lifestyles, rely on ducking and covering when danger approaches, rather than outrunning it. Both rabbits and hares practice hands-off parenting—the dads pay the young next to no attention, while the moms only see the little ones for about five minutes a day, to provide them with a swig of their very nutritious milk. For the European rabbits, *Oryctolagus cuniculus*, these five minutes per day are the only maternal attention the kittens ever get; after three weeks of speedy nursing the young are weaned, and their mother prepares to give birth to a new litter.

Because of their great fecundity and lack of fussiness when it comes to food, European rabbits (the ancestors of all domestic rabbit breeds) have posed problems in areas where they have been introduced by humans. Most notably in Australia, where their numbers grew from just a few dozen released in the mid-nineteenth century, to 10 billion in 1926, covering most of the continent in a gray carpet. Rabbits compete with native wildlife for food and habitat, their appetites clear the ground of plants, which leads to soil erosion, and their gnawing on tree bark destroys saplings. Interestingly, while down under rabbits are such pests that biological control was deployed against them (the numbers are down to a mere 200 million these days), the Red List of the IUCN classifies the species as . . . endangered. That's because in their native Iberian Peninsula, the European rabbit population is rapidly declining due to disease and habitat depletion. In turn, the loss of bunnies causes problems for a species that feeds almost exclusively on them—the endangered Iberian lynx, *Lynx pardinus*.

Meanwhile, rabbits themselves are herbivores—but what *The Tale of Peter Rabbit* and other classics fail to mention is the fact that bunnies are also coprophagous, or poop-eaters. While coprophagy is not uncommon among herbivores, rabbits need to consume their feces, otherwise they develop malnutrition. They have actually developed two kinds of poo: the first time the food goes through, it comes out the rear end as a "cecotrope"—a soft dropping that does not require chewing—and is reingested daily, directly from under the tail (which is why these soft pellets are not often seen). The second time the food makes it through the digestive system, it is expelled as the round, hard, nutrient-poor pellets that can be found scattered throughout rabbit homelands. This process ensures the rabbits get the most nutrition out of their vegetarian diet: it's so nice they eat it twice.

Face mites

Demodex folliculorum, Demodex brevis

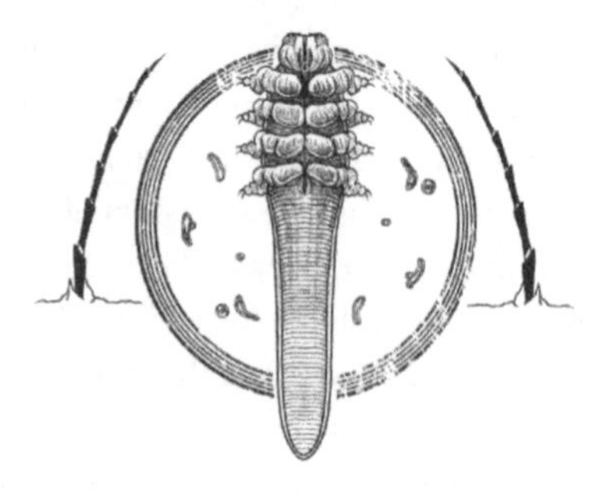

"You'll Never Walk Alone," sing Liverpool Football Club fans. It's more true than they might realize.

Face mites, *Demodex folliculorum* and *Demodex brevis*, are arachnids measuring a third of a millimeter, and, true to their name, they inhabit human faces. Don't run off to scour your forehead and nose quite yet, though; these little arthropods are a perfectly normal part of our local, bodily zoo—everyone's, not just Liverpool fans'. The worm-like mites have eight legs on the front of their body; *D. folliculorum* are a tad the bigger of the two species, and reside in groups in hair follicles, while the spindle-shaped *D. brevis* are solitary and live in sebaceous glands. Their favorite foods are sebum (skin oil) and epidermal cells, which they hold and eat with their spider-like mouthparts. To ensure the best dining spots, their residence of choice is in oily areas of the skin: nose, cheeks, forehead, and chin; they may, however, be found further south, on the chest or the genitals. The clue lies in the name, coined in 1843 by the English biologist Richard Owen, from the Greek *dēmós*, "lard," and *dex*, "boring worm," nicely summing up the looks and preferences of the mites.

Like a lot of other animals, these minibeasts are nocturnal, and rest during the day. While we are sleeping, they leave their shelters and go on a walkabout, at the leisurely pace of 16mm/h, looking for a mate. When one is found, sweet mite lovemaking takes place around our eyebrows, beards, or near the entry of other hair follicles. Females then retreat into the safety of sebaceous glands to lay eggs, which hatch within sixty hours. The larvae hatch into protonymphs and nymphs before reaching their adult stage, and, all in all, the tiny arachnids only live for 2–2 ½ weeks. Because they don't have anuses, face mites accumulate their waste material in their abdomens, and when they die and disintegrate inside a skin pore, a lifetime of feces is released in one messy go.

We share a truly special relationship with the two face mite species, being their exclusive hosts; they are the largest and most complex organisms of all those inhabiting our skin. In most cases, the presence of the mites is inoffensive, and, rather than being considered parasites, they are classed as commensal (from *com*, "together," and *mensa*, "table"), meaning that they benefit from living on our skin, but neither help nor harm us. However, given the opportunity, the innocent arachnids turn parasitic. On a healthy person, the immune system keeps their numbers in check, but people who are immunosuppressed tend to have more mites—even ten times as many as a healthy human. This increase in mite numbers is particularly acute for those suffering from rosacea (red, inflamed skin with swellings and visible blood vessels). In fact, rosacea is probably triggered by the bacteria contained in mite feces. To make matters worse, the stress associated with rosacea changes the chemical composition of the sebum, rendering it more nutritious for mites and helping their numbers grow even more—aggravating the condition further.

Face mites are very egalitarian—they have been found all over the world, on people of all ethnicities. The likelihood of carrying them increases with age; children are born mite-less and acquire them from adults. Remember that dreaded childhood auntie who always insisted on kissing you? Yep, she's to thank for your real-life cooties. In adults, infestation rates vary between 20 and 80 percent, reaching 100 percent in elderly people. Are you infested? Well, you mite be.

Giant panda

Ailuropoda melanoleuca

Giant pandas, *Ailuropoda melanoleuca*, are probably the most distinct and bizarre members of the bear family. Like other bears, the panda is classified as part of the carnivore order—however, it didn't seem to get the memo, as its diet is plant-based, consisting almost exclusively of bamboo. This food choice is troublesome for several reasons. First of all, despite the very specific vegetarian preference, the giant panda's digestive tract is still that of a carnivore, and it lacks the genes necessary for the complete processing of bamboo. Aided by gut microbes, giant pandas can digest only about 17 percent of what they consume. Thus, to meet their energetic needs, these bears require a high intake of food (up to 45 percent of body weight daily), which means that their life pretty much revolves around eating: they spend approximately fourteen hours each day foraging. While other bear species can ramp up their calorie intake from 8,000 to 20,000 kcal/day in times of need, the pandas only obtain 5,000 kcal/day (and spend 3,500 of it)—which does not allow them to store much fat for hibernation, pregnancy, or nursing.

With their minds continuously preoccupied by their next meal, it is no wonder that pandas are rarely in the mood for anything else. With the smallest average litter size of all bear species, giant pandas tend to have a low reproductive rate and are also notoriously difficult to breed in captivity. To begin with, it is incredibly tricky to fit potential mating around their schedules: the females experience only a single three-day period of sexual receptivity per year. The males, meanwhile, either lack sexual motivation, or are excessively aggressive. In captivity, the female who is ready for handy-pandy is presented with a selection of mates. If all the males fail to impress, the keepers resort to artificial insemination.

Despite the obstacles, the determination of conservationists from China and around the world has resulted in great successes in the panda-making department: as of 2016, the species is no longer classified as endangered. It is a good thing that the giant panda is the official emblem of conservation efforts, because its chances of becoming a symbol of great parenting are rather slim. The fathers take no interest in their offspring whatsoever—in fact, they never meet them. Meanwhile, the females make such nonchalant mothers that sometimes they accidentally sit on their babies and crush them to death. In their defense, compared to the 100kg mommy, the cubs are particularly tiny: they weigh around 120g on average—as much as a hamster. The 900-fold size difference means that giant pandas have, proportionally, the most minuscule young of all placental mammals.

Even though pandas often give birth to twins, they tend to adopt a cold-blooded one-child policy. The mother picks the stronger of the two cubs and focuses her attention on it, while ignoring the other one, which results in the death of the weakling. To be fair, the lack of nutrients in bamboo does make it nearly impossible for her to produce enough milk for two young. From an energy viewpoint, investing in one healthy cub is a better option than potentially losing both.

Thankfully for the species, captive pandas are assisted by a whole host of highly specialized nannies: zookeepers, vets, and researchers. One of the cubs may be taken into human care immediately after birth, leading the panda mom to believe that she only has a single baby. The cubs are then swapped several times a day, allowing them both to receive their mother's milk.

If mama panda fails to provide maternal care altogether, the cub is looked after by the keepers, while the female receives training in maternal behavior. Tutorials are conducted with the help of a toy panda doused in the real cub's urine, and recordings of baby panda's vocalizations. Sometimes keepers resort to milking the female, in the hope of maintaining milk production and reuniting the mom with her baby (if she eventually starts taking an interest in motherhood). Commitment to conservation is limitless: imagine being the milkmaid for a 100kg bear!

Giant prickly stick insect

Extatosoma tiaratum

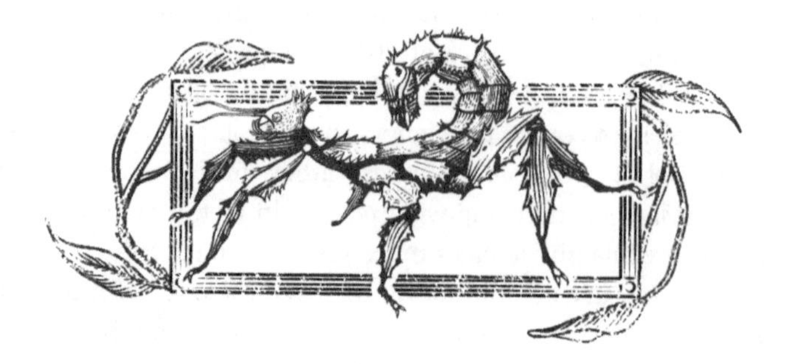

Ah, James Bond's gadgets! Supplied by the ingenious Q from the research and development division of the British Secret Service, they always get 007 out of trouble. And if Bond, with his arsenal of Q's brilliant inventions, had an animal equivalent, it would surely be the giant prickly stick insect, *Extatosoma tiaratum*. This leaf-eating Australian arthropod is so well kitted out when it comes to spy devices that it would put any agent to shame.

Like all members of their family, these stick insects employ visual camouflage to blend in with their surroundings. Also known as Australian walking sticks, they resemble foliage in shape. Based on their habitat, the stick insects' coloration is similar either to dried leaves, or—at higher altitudes—to lichen. Their disappearing act works well if the animal remains motionless for a long time, since moving around can render the carefully evolved costume redundant. However, a problem arises if the leaves they attempt to mimic are blowing in the wind, as remaining stationary against a mobile background would be just as revealing. Thankfully the giant prickly stick insects have figured out motion crypsis: when they sense stronger air movements, they start swaying and rocking on their legs to make themselves even more realistically leaf-like.

This spy tactic is employed by adults, but what must be the greatest feature of the Australian walking sticks' secret-agent toolkit is the fact that they come equipped for every stage of their life. Stick insects start off as eggs, flung by their moms from the trees on to the ground. By the way, stick-ladies don't resemble Bond girls (except in stunning good looks)—if a male is not around, rather than being a damsel in distress, the females simply produce eggs on their own, by parthenogenesis. Sure, there aren't as many, but any clutch is better than none. The eggs are highly attractive to ants, which pick them up and carry them to their nest, where they eat the lipid-rich outer layer and dispose of the rest in their waste piles. There, the eggs can take a few months to hatch—and when they do, the young stick insects look very much like ants, with orange heads, dark bodies, and swift movements. They even curl their abdomens to look more ant-like. Their myrmecomorphy (resembling an ant, also seen in some jumping spiders, see page 44) confuses visual predators, such as birds or reptiles. In this way, the hatchlings, called nymphs, are protected as they emerge on their brisk climb to their adult home: the tree canopy.

The nymphs only look like ants for the first few days of their lives—as they develop, they start resembling adult stick insects. Still, on their initial reconnaissance, they make use of other nifty characteristics: apart from donning an ant outfit, the hatchlings have the ability to glide (useful insurance for when a leg—or six—slips), and even to walk on water if they need to cross ponds.

When the walking sticks moult and acquire their adult features, it becomes apparent that the species is sexually dimorphic—that is, the sexes differ in looks and defensive strategies. The females are larger, reaching up to 20cm in length; they are thorny and broad-bodied, and, if in danger, they can draw blood by using their hind legs as pincers. The smaller males (up to 12cm in length) are slender; they have wings that can either be used to fly away from danger, or be flashed unexpectedly in a warning display. Additionally, when threatened, the stick insects will curl up their prickly abdomens, pretending to be scorpions; they also make clicking sounds. To top it all, they can generate a chemical secretion from their mouthparts, which smells, somewhat surprisingly, like . . . toffee. We can only hope that in the next instalment of the James Bond franchise, 007 will be able to walk on water, glide from tree canopies and emit a faint caramel odor when confronted with villains.

Iwasaki's snail-eater

Pareas iwasakii

Escargots, anyone? There is a surprisingly high number of snakes who would raise their hands in response—that is, if they had hands.

Mollusc-eating snakes belong to a group called goo-eaters—yes, this is the technical term used by herpetologists—that feed on all things slimy: snails, slugs, worms, and occasionally amphibian eggs. While eating a slug is easy if one doesn't mind the mucus (and goo-eaters will diligently wipe their mouths on the ground after a slug meal), snails pose a problem—they come pre-packed. Some snakes deal with shelled items, such as eggs, by swallowing them whole in an enormous, eye-popping gulp, crushing the shells internally, and then spitting them out. Goo-eaters are much more sophisticated: they use a range of adaptations to extricate the snail without breaking the shell—no mean feat for a creature with no fingers.

Some goo-eaters, such as the clouded snake (*Sibon nebulatus*) from the American tropics, use the "snag-and-drag" technique of snail extraction. The method is not dissimilar to what a person does with escargots in a French restaurant, except, of course, nobody cooks the snails in garlic butter for a snake. The reptile grabs the snail by the head with its mouth, gets a firm grip on it, and then drags it along the ground until it finds an appropriate object to snag the mollusc on—for instance, a branch or a pointy rock. This snagging device is the equivalent of a brasserie's snail tongs, used to keep the food in place. The snake then anchors itself firmly to the substrate with its tail, flexes its muscles and uses its snail fork—the nimble but sturdy jaw—to pull the morsel out. Voilà!

On the other side of the world, the snail-eating pareid snakes from Southeast Asia, such as Iwasaki's snail-eater, *Pareas iwasakii*, from Japan, have taken the refining of their snail forks one step further. To extract their asymmetric prey, they have developed asymmetric mandibles.

Like people, snails come in a right-handed (dextral) and left-handed (sinistral) variety—although in the case of the molluscs, the handedness refers to the direction in which their shells coil. The direction of the coil affects the positioning of internal organs, including genitals, which leads to reproductive incompatibility between dextral and sinistral snails. As a consequence, either the right- or the left-hand type dominates in an area, simply because finding a matching mate results in even more snails with the same coil direction. Since right-handed snails are prevalent around the world, it makes sense for a snake that specializes in eating them to come appropriately prepared. Indeed, Iwasaki's snail-eater does just that: it has twenty-six teeth on its right mandible, and only eighteen on the left. The snake is able to move these two jaw bones independently, and, after grabbing on to a snail, can niftily tease it out of the shell by alternately retracting the right and left mandible.

While this asymmetric mechanism is super-efficient for eating dextral snails, the snakes find it extremely difficult to grab a sinistral snail—which means that in areas with Iwasaki's snail-eaters, left-handed snails are at an advantage. Because the left–right reversal is determined by a single gene, the predation on right-handed snails acts as an evolutionary driver, speeding up the proportional increase of left-handed snails. In fact, predation on right-handed snails by pareid snakes has contributed to an extraordinarily high diversity of left-handed snails in Southeast Asia.

Iwasaki's snail-eaters, who rarely come across sinistral snails, don't know how to extract them. However, other pareid snake species who are more likely to encounter sinistral snails are better at assessing whether the potential meal coils the right way, and can adjust their approach to save themselves time and extraction effort. It all goes to show that there is more than one way to shell a snail.

Jumping spider

Toxeus magnus

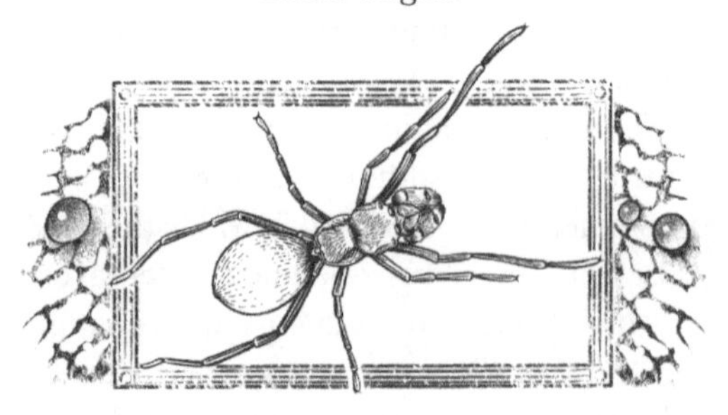

Jumping spiders, or salticids, are the most numerous spider family, comprising over 6,000 (13 percent) of all spider species. They are generally small, with a body length of between 1 and 25mm. They are also the cutest of all spiders, due to their large, front-facing, staring principal eyes, resembling those of puppies. The eyes are not just there to endear them to their opponents—the vision of jumping spiders is among the most acute of all invertebrates. Apart from the two principal eyes, they also have six others, including a pair located, quite literally, at the back of their head, giving them a 360-degree field of vision.

With such top-notch equipment, salticids are incredibly competent diurnal predators. They don't build webs for trapping their prey; instead, they use a variety of ambush and stalking techniques, and pounce on their victims. However, when they jump—and they jump about fifty times their body length—they tether a piece of silk to the substrate as a safety line, just in case. At other times they use the silk filament to lower themselves on to their unsuspecting victim. They may also follow prey over a complex path, persisting even after temporarily losing sight of it.

Some salticids have a rather difficult relationship with other spiders—they can enter the webs of web-building species, such as orb-weavers, and steal the ensnared insects. Why stop there, though? These cunning hunters will also slowly walk on top of a web, and make small vibrations with their legs and palps to mimic the movement of a trapped insect. The web owner eagerly shows up for dinner—only to become the salticid's main course.

It is not only the web-building spiders who are consumed—other jumping spiders will feature on the menu as well. Because of that, some salticid species developed a permanent Halloween costume, mimicking other invertebrates, such as ants, beetles, or wasps. While certain jumping spiders are specialized ant-hunters, myrmecomorphy, or looking like an ant (also seen in the giant prickly stick insect, page 40), generally helps to avoid predation: the ants' powerful bites and chemical defenses make them dangerous prey items. These spiders' mimicry is highly convincing—they have a "false waist" that makes them look more ant-ish, and wave their front pair of legs to resemble antennae. However, one such ant-mimicking spider, *Toxeus magnus* from Taiwan, is special for a completely different reason—it feeds its young milk.

Okay, so the milk might not be exactly the same as mammalian milk, as arachnids don't have mini-nipples or lips for suckling. But spider milk is a highly nutritious substance, containing four times the protein of cow's milk. Even though spider moms don't bring any extra food into the nests for the first twenty days, the hatchlings more than triple in size over this time.

The yellowish, wholesome secretion comes out from the mother's epigastric furrow (an opening at the base of her abdomen, used for laying eggs). At first, she deposits the droplets inside the nest for newly hatched spiderlings to drink; later, when the youngsters are a week old, they suckle straight from her belly. Hatchlings deprived of mother's milk from day one don't survive, although if the milk is withdrawn later in their lives, at three weeks of age, they have a higher chance of making it through. Doting mothers provide milk for their brood well after they have reached maturity, although the sons are cut off and sent packing much earlier than the daughters, probably to avoid inbreeding. While the offspring stay with the mom, she cleans and repairs the nest, reducing the risk of her family getting parasites. Nursing *and* cleaning adult offspring? The salticids take devoted motherhood to new extremes.

Millipedes

class Diplopoda

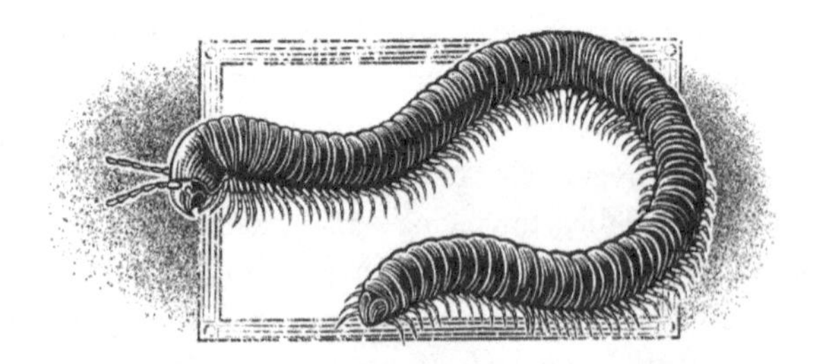

The leggiest creepy-crawlies of the animal kingdom are arthropods belonging to the subphylum Myriapoda: millipedes and centipedes, meaning "thousand-footed" and "hundred-footed" respectively. The flattened, carnivorous centipedes actually exceed limb expectations, with leg numbers varying between 30 and 382 depending on species (though, strangely, centipedes always have an odd number of pairs). In contrast, the docile, detritivorous millipedes (12,000 species in class Diplopoda) were thought to trail behind, never quite hitting the thousand-leg target. Perhaps the latter ought to have been called bicentipedes instead, since their body segments are fused in pairs, resulting in double the number of legs per segment compared to centipedes. However, in 2021, millipede honor was finally redeemed with the discovery and description of *Eumillipes persephone*, a small, Australian species boasting a staggering 1,306 legs—more than any other creature on Earth.

As millipedes moult, the number of their body sections increases—in a process called anamorphosis. Their multi-segmented bodies—along with all the legs—help them burrow more efficiently. In some species, the total sum of segments is fixed (teloanamorphosis), but others bear an uncanny resemblance to the game of "Snake" on a vintage Nokia phone: the longer they live, the more segments they get (euanamorphosis). As a taxonomic group, millipedes are even more ancient than the Nokia 3310. With the fossil of *Pneumodesmus newmani*, found in Scotland, dating back to the Early Devonian period, some 414 million years ago, millipedes hold the record for the oldest air-breathing terrestrial animals.

The longest millipede, the African shongololo, *Archispirostreptus gigas*, reaches over 33cm; the shortest ones are only a couple of millimeters long. Body shapes vary from the soft and tufted bristly millipedes, to the oval, roly-poly pill millipedes, to the long, worm-like rest. Quite appropriately, the name of one species, *Crurifarcimen vagans*, literally translates as "wandering leg sausage."

The leg sausages, though slow-moving and generally keeping themselves to themselves, are not consumed as often as one might expect from their name. While they are preyed upon by insects, birds, reptiles, amphibians, and mammals, many species protect themselves by secreting unpleasant defensive chemicals. These include aldehydes, quinones, chlorine, iodine, or hydrogen cyanide, and can be toxic, irritant, or sedative for potential predators. Toxicity often comes with a color warning (aposematism), so although most millipedes are black or brown, some sport a shocking-pink or bright red display.

The repellent secretions may be a bit of a double-edged sword. Even though they are meant to deter, some mammals, such as meerkats, capuchin monkeys or lemurs, will nibble and prod millipedes on purpose, prompting them to release the toxic substances. They then proceed to anoint themselves with a mix of saliva and the millipede product, as if using a multi-legged bottle of lotion. Most animals apply millipede secretions as medication or insect repellent. However, there is a chance that some, such as the black lemurs, chew on them just to get intoxicated. Red-fronted lemurs add insult to injury by concentrating the rubbing efforts primarily on their bottoms, genitals and tails, presumably to get rid of gastrointestinal parasites.

Not all animals have such a distasteful relationship with the leggy arthropods. Some millipede species are myrmecophiles—they form close associations with ants. Army ants in particular appear to be very fond of their millipede companions, who, by feeding on the dirt, mold, and organic debris in the ant nests, provide a free cleaning service. When army ants move to a new locum, their millipede orderlies travel as part of the column; if they lag behind, they are still able to catch up by following the ants' chemical trail. Sometimes they are even carried by worker ants—numerous legs clearly don't guarantee a fast enough walking pace.

Mole salamanders

Ambystoma spp.

"A woman without a man is like a fish without a bicycle" was the 1970s feminist slogan coined by Irina Dunn. But perhaps a better analogy would be that of the all-female mole salamander lineages, *Ambystoma*, who have made it through for 5 million years without males.

Ambystoma are a genus of salamanders from North America, containing thirty-two species, among them the famous axolotl. Thirty-two species? Well, sort of. While we tend to think of quadrupeds as simple beings in terms of sex lives—a mother, a father, a few children, all sticking to their own kind like a good species should—the mole salamanders spoil it all. They are an absolute nightmare for taxonomists and evolutionary biologists, because the American Great Lakes area is inhabited by a whole female-only lineage.

The first question that pops to mind is: how could they possibly have survived? Sexual reproduction is pretty much ubiquitous across the animal kingdom, and its biggest advantage is genetic recombination—the mixing of different individuals' genetic material to produce a diverse population. Diversity provides more scope for natural selection, and when a disaster strikes, chances are at least some of the individuals will be able to survive and adapt. That said, there are several species that reproduce outside of the classic, sexual route; some animals, for example, reproduce asexually, creating copies of themselves without any input from another organism. The all-female *Ambystoma* are not quite that, though—they are unisexual.

Unisexual vertebrates—of which there are eighty or so species—employ one of three reproductive modes. There is parthenogenesis, where eggs develop without fertilization; there is gynogenesis—very similar, but a sperm cell is needed to activate the egg development (though its genetic material is not contributing to the offspring); and hybridogenesis, where an egg is fertilized, but male genes are not passed on to subsequent generations. Unisexual *Ambystoma* are gynogenetic; they need sperm to kick-start the egg production. To do that, they resort to kleptogenesis: stealing sperm from males of four salamander species living in the same area. Salamander males usually deposit little sperm parcels that are used by females for fertilization—however, the unisexual *Ambystoma* ladies take the sperm to stimulate their reproduction, and then discard it, preferring to produce clones of themselves. If this weren't complicated enough, sometimes they will incorporate the genes from the males into the offspring, in a genetic equivalent of pick-and-mix—leading to very complex genomes.

Humans receive two sets of genes, one from the mother, one from the father, making us diploid. But unisexual mole salamanders can be triploid, tetraploid, or even pentaploid. And they can combine genes from several species of neighborhood males in one egg, which is a bit like finding out that your DNA contains the genes from Mom and Dad, but also a gorilla, a chimp and an orangutan. The mix-and-match approach to genomes completely ruins the traditional species concept.

How can we know that all those unisexual salamanders don't in fact belong to four separate species? The clue is in the mitochondrial DNA, which is only inherited from moms, thus making it possible to neatly determine the maternal lineage. In the case of mole salamanders, the mitochondrial DNA is similar across unisexual *Ambystoma*, but different from all four "parental" species, which pitched in their genetic material at various points in evolutionary history. Interestingly, the unisexual salamanders are very successful, and outnumber their sexual counterparts by as much as two to one in some populations. Mole salamanders: the original exponents of girl power?

Mountain tree shrew

Tupaia montana

Tree shrews are not shrews—nor do all of them live in trees. They are small, brownish mammals, native to Southeast Asia, and they resemble pointy-nosed squirrels. They are taxonomically unique, sitting in their own order (Scandentia), and are one of the primates' closest relatives. They also boast the biggest brain-to-body ratio among mammals—yes, bigger than that of humans. Because of this—as well as a longer life span (9–12 years) than laboratory rodents—tree shrews have been used in biomedical research investigating psychosocial stress, myopia, viral hepatitis and Alzheimer's. In the wild, however, tree shrews have different responsibilities. One species, the mountain tree shrew from Borneo, *Tupaia montana*, enjoys a very special relationship with plants.

Most of the time, plant–animal relations are rather one-sided, with the animal eating the plant. However, sometimes the tables turn. Plants already have a tough time—they need elements such as nitrogen or phosphorus to survive, but being immobile makes it difficult to obtain these, particularly if one is stuck in nutrient-poor, tropical soil. It is no surprise that some plants develop inventive solutions to supplement their nutrition—for example, by befriending symbiotic bacteria or fungi. To improve their chances of getting a square meal, a number of plants resort to a trick usually associated with animals: carnivory. Venus flytraps use a snap-trap mechanism to lock flies between their leaves, sundews stick insects to their gluey surfaces, and pitcher plants use pitfall traps made of modified leaves to capture unsuspecting insects. But several species of pitcher plants from the genus *Nepenthes*, native to Borneo, enhance their carnivorous diets with something quite unexpected.

Rather than having the usual insect-trapping, slippery, lightweight pitchers, the giant Bornean pitcher plant, *Nepenthes rajah*, developed much bigger, broader, and more robust containers. What's more, the "lids" of its pitchers produce rich, buttery nectar. What for? Does this species hunt bigger game than insects? Sometimes it does—*N. rajah* is one of three pitcher plant species known to have caught mammalian prey; it can also capture small amphibians and reptiles. Still, this is not the main reason for its peculiar adaptations.

The Bornean giant has an interesting deal with the mountain tree shrew. Tree shrews are omnivores, feeding mainly on arthropods and fruit, but, like humans, they are partial to a bit of sweetness. To access the sweet, buttery goo that the pitcher plant exudes, they have to hop on to the pitchers (which are sturdy enough to support an animal weighing over 150g). After it has fed, the tree shrew deposits a dropping. The poop falls right inside the pitcher, which now serves as a little tree shrew toilet. In a feat of adaptability, the plant has evolved to match exactly the dimensions of the tree shrew's bottom. It is very secure, wide-brimmed and comfortable; and the shape of the orifice and the orientation of the lid force the tree shrew to sit astride the pitcher—just to make sure that it aims properly. Fortunately for the plants, it so happens that tree shrews produce very nutritious fertilizer; that's because food passes through their guts very quickly, in under an hour, and thus fewer nutrients are extracted from the diet. The transaction is so successful that another species with the same approach, Low's pitcher plant (*N. lowii*), gets between 57 and 100 percent of its nitrogen from tree shrew droppings.

Tree shrews have their favorite pitcher pit stops, which they mark using scent glands. They mainly visit during the day—at night, the plants are the potties for rodents, the summit rats, *Rattus baluensis*. As an additional signal, pitcher plants advertise these jungle service stations by providing special visual cues, highlighting the undersides of the lids with much brighter, contrasting colors in the blue and green wavebands visible to tree shrews (something that insect-feeding pitcher plants don't do).

It really goes to show that one creature's turd is another's treasure.

Mudskippers

subfamily Oxudercinae

The quote "the word impossible is not in my dictionary" is attributed to Napoleon Bonaparte—although it also seems to be the motto of mudskippers, who are not deterred from climbing trees and scaling rocks by the mere fact that they are fish.

Mudskippers are a group of thirty-two mangrove-dwelling tropical species in the subfamily Oxudercinae, related to gobies. While some bear rather traditional scientific names—for instance, the giant mudskipper, *Periophthalmodon schlosseri*, celebrates the Dutch physician and naturalist Johann Albert Schlosser, and the pug-headed mudskipper, *P. freycineti*, the French explorer Louis de Freycinet—the New Guinea slender mudskipper, *Zappa confluentus*, honors the musician Frank Zappa "for his articulate and sagacious defense of the First Amendment of the US Constitution."

Frank Zappa stood up for freedom of expression, and mudskippers are equally tenacious; they won't let anything hold them back. Of all amphibious fish, they are the most land-adapted, and may spend 90 percent of their time out of water. No lungs? No problem; they breathe either via water stored in their large gill chambers, or using cutaneous respiration—directly through the lining of their mouths and throats. Anything goes, provided they keep moist.

Suffering from dehydration? Mudskippers have a behavioral solution: they periodically roll in mud to keep their bodies lubricated; plus they are rarely further than a minute away from water. When not active, they take refuge in water-filled burrows, in which they also breed. Because the mud in which they build their lodgings is anoxic, and during high tide it is covered by water for a good few hours, mudskippers such as *Scartelaos histophorus* stock up on oxygen in advance. Their burrows are J-shaped, with an upturned portion not connected to the surface, which the fish use to create an air phase. They do so by inflating their mouth chambers with fresh air from the outside, then popping inside the house and releasing it there—up to fifteen times per minute—in preparation for the tide.

No feet to walk on land? Pah, these fish make do with fins. Their pectoral fins have a joint analogous to an elbow, used to propel the mudskipper forward in a motion resembling a person walking on crutches. While the pectoral fins do the lifting, the body is supported by a pair of pelvic fins, which are either fused, like a little suction cup, or unfused and gripping, allowing some mudskipper species to climb trees. These talented fish are also capable of hopping across water without getting their fins properly wet; they use their tails to bounce on the water's surface before reaching a tree, rock or another bit of dry land on which they perch, in the most un-fish-like fashion.

When not channeling Napoleon or Zappa, mudskippers embrace their inner Donald Trump and build walls. Many species, like *Boleophthalmus boddarti*, are hugely territorial, and will demarcate their polygonal dominions with a 3–4cm-high partition. They construct it by carrying gobbets of mud in their mouths, and devote about 5 percent of their daily activities to building and repair work. Thus delineated, these territories are also foraging grounds for nutritious algae.

Other, carnivorous, mudskipper species face a different setback—the absence of a tongue. Fish are not huge in the tongue department anyway (see Tongue-eating louse, page 146, for replacement options), and generally just suck in the water that contains food. On land, however, a muscular tongue that pushes morsels to the back of the mouth for swallowing is a better option than suction. But what mudskippers lack on the anatomy front, they compensate for with innovation. They fill their mouths with water, move it forward to grab the food, and then suck it back to swallow—mirroring the action of tongued animals.

Looking at various bits of mudskipper anatomy sheds light on vertebrate transition from water to land—and it seems a can-do attitude is just as important as the physical advantages land-based animals enjoy.

Naked mole-rat

Heterocephalus glaber

Naked mole-rats, *Heterocephalus glaber*, are small rodents—but, judging by their physiology, behavior, and ecology, their aspirations are far beyond what is expected of a mouse-sized animal. This subterranean, East African species is perhaps the least mammalian of all mammals.

Its alternative name, the sand puppy, is too kind—while they do live in soil, naked mole-rats look nothing like puppies. Appearance-wise, this rodent joins the proud team of creatures who could be mistaken for a human penis; hairless, loose-skinned, and longish, it is the only mammal in the ranks of this noble club. The most obvious aspect differentiating naked mole-rats from naked man-parts is the presence of bucked, ever-growing teeth used for digging tunnels. The teeth are positioned outside the lips, which prevents dirt from getting inside the mouth.

Living in poorly ventilated, underground mazes puts unique adaptive pressures on sand puppies. Rather than maintaining a steady body temperature, like a good mammal, they are thermo-conformers: their body temperature, like that of reptiles, changes based on the environment and is altered behaviorally, for instance by moving to cooler areas, or huddling for warmth. But huddling comes at a risk—subterranean chambers are low in oxygen, and animals sleeping at the bottom of the body pile (mole-rat families reach 300 members) can lose consciousness. Still, overall they are extraordinarily well adapted to hypoxic conditions, and may survive without oxygen for up to thirty minutes. This ability is linked to another superpower—naked mole-rats are impervious to certain types of pain, because they lack the substances involved in pain transmission.

Sand puppies seem to have forgotten that, as small mammals, they ought to die after only a few years. Instead, they can live to be thirty-three—longer than tigers or polar bears. What's more, they reach that old age in good health, with minimal signs of senescence. They are also resistant to cancer.

Perhaps the most interesting feature of these rodents is their family life. Like honeybees, ants, and a handful of other invertebrate species, naked mole-rats are eusocial (from the Greek *eu*, "good"), or "truly social:" they live in supremely cooperative families with a division of labor—though it is far from a democratic society. The queen, twice the size of a regular mole-rat, physically intimidates the rest of her colony into submission. She is the only breeding female and mates with a handful of reproductive males. The rest of the animals are non-reproductive and help run the colony by providing food, excavating tunnels or tending to the young. After the queen's death, a bloody battle determines her successor—the most ferocious female takes over the reign. Meanwhile, even though the species is tolerant to inbreeding (seeing that the colony is all related and the queen is likely to bear the young of her brother or son), there are some males who belong to a specialized "disperser" morph. Dispersers have more fat tissue, a nocturnal activity pattern and a tendency to leave the burrow; they travel up to 2km in search of a new colony with mating opportunities.

Naked mole-rat queens can bear one of the biggest litters among mammals, up to twenty-eight pups at a time. Still, looking after them is not too taxing—after a month of nursing, Mom hands over the youngsters to their older siblings, who feed them (in true older-sibling fashion) meals of poo. Since sand puppies eat plant roots and tubers, which sometimes require digesting multiple times to get the most out of their nutrient content, coprophagy, or eating feces, is nothing unusual for them (other species do it too, see European rabbit, page 34). The queen also uses dung dinners as a means of steering the colony babysitters—when she is pregnant, the hormones in her droppings will kindle maternal instincts in the rest of the group eating her fecal matter. It appears a combination of bullying and crap meals is a highly effective way of ruling an underground organization.

Pangolins

order Pholidota

The shy, unassuming pangolin shot to international notoriety in 2020, when researchers from Guangzhou suggested it might be the intermediate host of SARS-CoV-2, and that China's extensive and illicit wildlife trade is the reason behind the COVID-19 pandemic. Even though, after comparing coronaviruses from pangolins and people, the former were acquitted, wildlife trade for food and traditional medicine is still a major source of animal-borne diseases—and one of the biggest drivers of biodiversity loss. Pangolins hold the sad record of being the most trafficked creatures in the world, with an estimated million individuals traded globally between 2000 and 2013, and the numbers probably rising.

Pangolins are the sole members of the order Pholidota, containing four Asian and four African species; their body lengths range from the 40cm long-tailed pangolin to the 140cm giant pangolin. They are the only mammals covered in keratin-based scales, which, being one of the key ingredients of traditional Asian medicine, is the main reason for their demise. Although, from a biochemical perspective, pangolins are akin to being covered in toenails, the scales are believed to increase lactation, reduce swelling, and treat rheumatism and asthma. On top of that, pangolin meat is consumed as a delicacy and a status symbol.

Across Africa, pangolins are also used for medicinal, spiritual, and culinary purposes, although they are increasingly exported to China and Vietnam due to the plummeting numbers there. The closer to Southeast Asia, the more threatened the species are: Chinese, Philippine, and Malayan pangolins are classed as critically endangered, the Indian pangolin and two African species as endangered, and the two remaining African species as vulnerable. Probably not for long, though, since African pangolins are being exported on a massive scale, with Nigeria emerging as one of the biggest trafficking hubs. Nigeria alone saw seizures of scales from 800,000 pangolins between 2010 and 2021 (and these were just the ones that were caught). To make matters even worse, the historical demand for pangolin leather in the US, for the production of patterned cowboy boots, has further contributed to their decline.

But what are pangolins themselves like? Under their armor, their long snouts bear a look of puzzled, slightly nervous but kind-hearted innocence. When threatened, these nocturnal, solitary and rather introverted animals roll into a ball—a process called volvation in science-speak, and *pengguling* in Malay (hence their common name). The strong and sharp scales protect them from predators—even as sizeable as lions or hyenas—but, of course, not from poachers, who simply throw the scaly footballs into a bounty bag. Additionally, the African ground pangolin is threatened by electrocution from electric fences built around ranches—electricity is not impressed by volvation.

When not rolling up, pangolins forage. They feed almost exclusively on ants and termites, which they dig up using the massive claws of their forefeet, and scoop out with their long, thin tongues covered in sticky saliva. If you thought pangolins looked weird on the outside, they are even stranger on the inside. Their tongue reaches 40cm in length, and, counterintuitively, it does not roll up like a tape measure, but passes through the torso and is anchored at the pelvis. Because they lack teeth, pangolins grind up their meals in the same way birds do, using a kind of gizzard. Inside this thick, muscular part of the stomach, there are gastrolites—small stones that help to macerate food. On top of that, Malay and Chinese pangolins have scales on the outside *and* the inside, since their stomachs are lined with "pyloric teeth"—scale-like, keratinised spines, for even better digestion.

Because there are few studies of pangolin ecology, and none of the species breeds well in captivity, their private lives are still shrouded in mystery. With more research and better awareness of these aloof oddballs, we can hope that they will be better appreciated as living beings instead of piles of scales.

Pseudoscorpion

Paratemnoides nidificator

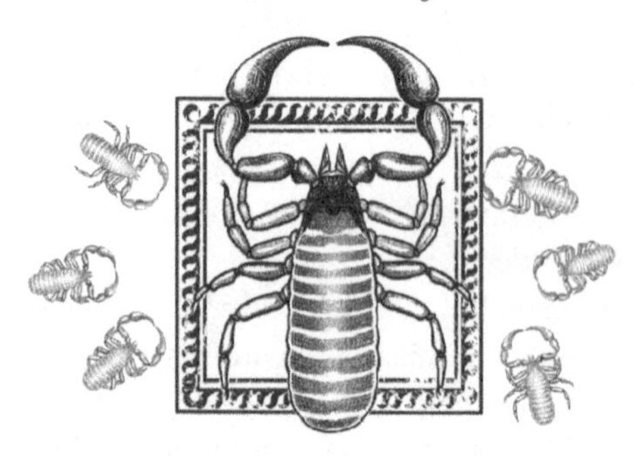

Due to their pronounced pincers, pseudoscorpions look a bit like teeny versions of their bigger and better-known cousins, scorpions. Like their other cousins, spiders, they are able to produce silk. Pseudoscorpions are petite (2–8mm long) arachnids; they live under tree bark, stones, or leaf litter, but may also be found in people's homes, especially in old books, gaining them the nickname "book scorpions." If you see one—don't panic, they are not harmful to humans. On the contrary, they are a fashionista's friend, since they eat the larvae of clothes moths.

There are over 3,400 described species of pseudoscorpions; they are very widespread, inhabiting areas from Canada to Australia, with the highest diversity occurring in the tropics. Being very little, they would have a hard time covering long distances—however, these tiny arachnids have transportation figured out: they hitch lifts. The technical term for such hitchhiking is phoresy—smaller animals attach themselves to larger ones for a free ride. In the case of pseudoscorpions, the transporters are usually insects, such as beetles, bugs, or earwigs, and the first-class seats are on their legs and antennae. Some itinerants travel alone, some in groups of two to seven. For the insect, there are advantages to carrying passengers: pseudoscorpions—in a dining-car experience of sorts—will happily feed on pesky parasitic mites. Then again, the mini stowaways might hang on to the insects until the very end; when the transport dies, the remnants become a scrumptious final-destination buffet.

Most pseudoscorpions are rather antisocial, and will generally pick a fight if they come across a trespassing rival. Such a fight ends with a meal: the loser becomes dinner. Book scorpions are voracious predators; they will hunt invertebrates of all sizes—and will stoop to cannibalism. Still, some species, such as the *Paratemnoides nidificator* from South America, are sociable, live and hunt together, and share food. *P. nidificator* occupy communal nests of over 200 individuals, consisting of both adults and younglings called nymphs. The nymphs work together to build communal moulting chambers, while adult females construct individual brood chambers, where they rest with the eggs—both types of compartments are made of silk, which sounds rather comfortable. Whether solitary or social, pseudoscorpions make very good mothers, looking after the embryos and nymphs by feeding and grooming them. The female stays in the brood chamber until her nymphs emerge, and then the little family group leaves the silk premises to feed on prey captured by other adults in the colony. Males and non-reproductive females share their food with younger siblings and offspring.

While individual animals usually consume prey smaller than, or as big as, themselves, hunting collaboratively allows *P. nidificator* to take down insects several times their size. Hunting in packs means more capacity for chasing, attacking, and subduing prey, such as beetles, stinkbugs, ants, and spiders. Pseudoscorpions use their pincers to hold down the victims' appendages or inject venom into their joints. It's not a case of all pincers on deck—while some animals attack, some merely observe, and others, the profiteers, only show up for the food. Still, the final feast is peaceful and orderly: the hunters and observers feed first—though they will make way for any hungry nymphs—and the idle profiteers get the scraps.

Things get interesting—and more drastic—in times of hunger. With no prey to feed her nymphs, mother pseudoscorpion makes the ultimate sacrifice. She leaves the nest, raises her pincer-like pedipalps, and invites the young to feed on . . . her. She will stand, calm and motionless, as the nymphs attack her. They focus on the joints, where the exoskeleton is the thinnest, and literally suck her dry. When she is just an empty shell, she is cast aside, and the young nymphs, energized by the meal, venture out to hunt cooperatively. This phenomenon, known as matriphagy, is thought to serve as a means of ensuring nymph survival and reducing cannibalism among the offspring. It certainly puts pseudoscorpion moms among the most devoted and sacrificial parents in the animal kingdom.

Red-eyed tree frog

Agalychnis callidryas

The red-eyed tree frog, *Agalychnis callidryas*, from the tropical forests of Central America, is a real stunner. Its slender body is bright green with striped blue-and-yellow flanks, its belly white, its feet a vibrant orange and the eyes a striking red. The frog's species name, *callidryas*, reflects its good looks: in Greek, *kalos* means "beautiful," and *dryas*, a "dryad," a forest spirit.

Bright colors on a tropical frog—surely this means it's poisonous? Actually, no. The vivid coloration is not aposematic (a warning); instead, the red-eyed tree frog uses camouflage to avoid being noticed. Although it might seem that the only place where this sort of camouflage could work is a clown convention, the tree frog can make itself almost invisible when resting. It does so by covering its blue sides with its legs, tucking its orange feet under its body, and closing its red eyes. When approached by a predator, it will suddenly open its bright eyes to startle the opponent and buy a few seconds of surprise to get away—an example of deimatic, or frightening, behavior.

How do the tree frogs know when danger approaches? These amphibians are nocturnal, and rest during the day. Their eyelid equivalent, the nictitating membrane, is semi-transparent—and, covered in a subtle gold pattern resembling a Middle Eastern veil, is as beautiful as the rest of the dryad itself. These specialized eyelids allow the frogs to sense changes in the light and react when unwanted visitors are near.

However, it is not just the adult frogs that have an inbuilt alarm. Little red-eyed tree frogs can sense danger—and react to it accordingly—while still inside their eggs. Such an early-warning system is dictated by the frogs' life history. The tree frogs lay jelly-covered eggs on vegetation above ponds—when the young hatch, they fall into the water to begin their lives as tadpoles. However, before they grow up, they are faced with a range of potential threats—first from the land or air, then in the water. If there are no disturbances around the leafy nursery, it makes most sense for the eggs to develop on the leaf for as long as possible, to avoid being eaten by fish or freshwater shrimps in the pond. But what happens when peril strikes while the frogs are still in the eggs, stuck to a leaf?

The frogs' solution to this problem is to speed up the hatching process. When undisturbed, the eggs will take about a week to hatch; however, if the clutch is attacked by a snake or a wasp, the embryos are capable of hatching as early as day four. The anti-predator response happens at a breakneck speed: it is triggered on average sixteen seconds after the snake attack commences, and within five minutes all embryos are hatched (or eaten). The emergence is prompted by vibration across the clutch caused by the predation.

Premature hatching can save lives—but, if it's a false alarm, it can be costly, as the risk of predation in the water is higher for less-developed tadpoles. Fortunately, red-eyed tree frog embryos are able to assess whether the vibrations transmitted across the gelatinous eggs are generated by a snake, or, say, rainfall. They use the duration, intervals, and frequency of the vibrations to decide whether they should hatch or not—and they rarely hatch prematurely during rainstorms. The beautiful dryads are smart as well as good-looking.

Saharan silver ant

Cataglyphis bombycina

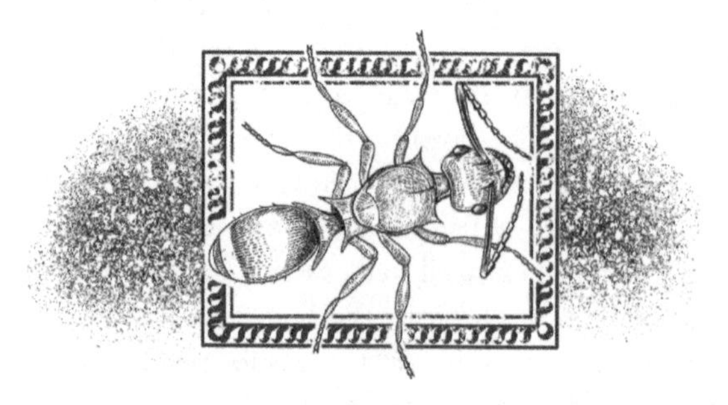

The Sahara is a punishing place to live. During the hottest hours of the day, surface temperatures easily exceed 60°C; water is extremely scarce, and food hard to come by. Loose sand impedes walking. Sparse vegetation, scorching sun and a clear sky make shade a rarity. Under these extreme conditions, exposed to overheating and desiccation, most insects would shrivel up and die.

If they have any sense, desert-dwelling animals will protect themselves by spending their days in hiding—in burrows or under rocks—and emerge during the cooler hours of night. Not so the Saharan silver ant, *Cataglyphis bombycina*. No—these ants are active in the hottest, most dangerous time of the day, when no other creature dares to venture into the desert.

Silver ants exploit the wussiness of other animals in two ways. Firstly, they are thermophilic (heat-loving) scavengers, which means they feed on the corpses of arthropods that succumb to heat exhaustion. Secondly, not wanting to be consumed themselves, they leave their nests when their main predator, Duméril's fringe-fingered lizard, *Acanthodactylus dumerilii*, retreats into a burrow after deciding that it's too hot to stay outside.

The ants are active on the surface only for some ten minutes each day, emerging around 1 p.m., when the sun is at its most grueling. The colony's foragers will appear for just a few minutes in a dramatic outburst, when the heat at ant height reaches 46.5°C. Silver ants are able to withstand a maximum body temperature of 53.6°C, which leaves a very short window of foraging opportunity before they themselves are fried. They may climb a stone or a dry bit of grass to cool down a little, as the temperatures are lower even a few centimeters above the ground. However, any delays in getting back to the nest carry lethal consequences, thus silver ants need to travel fast and navigate well.

And they most certainly do: Saharan silver ants are among the fastest land creatures alive, capable of covering 85.5cm/s. Their walking speed is comparable to that of humans, except, of course, they are less than a centimeter long. In terms of body size, a 180cm-tall person would have to run at 720km/h to achieve the same feat. A key to the silver ants' success is their stride frequency of over forty steps per second, allowing them to outpace their longer-legged cousins, Saharan desert ants, *C. fortis*. Moreover, to lengthen their stride and help them scale tricky sand dunes, their gait resembles the galloping of quadrupeds.

Both ant species show impressive navigational skills. Rather than retracing their meandering foraging steps to return to the nest, they are able to reach it in a straight line. To arrive home this efficiently, they need two pieces of information: what direction they ventured out in, and how far they have walked. To calculate the former, they probably use a skylight compass, measuring the light polarization of the sun. The latter they figure out with their inbuilt pedometers. The reliance on step-counters was studied in *C. fortis*—when stilts were attached to their legs, making their steps longer, the ants overshot the distance to their nests.

Saharan silver ants have more cooling tricks up their sleeve. Their bodies synthesize and accumulate protective heat shock proteins—but, unlike most animals, they do it pre-emptively, rather than as a response to the heat. Producing these proteins prior to heat exposure allows the insects to reach their feverish body temperatures and reduce any damage.

Another clue behind the silver ants' cooling success lies in their name: these animals are covered in reflective hairs with a triangular cross-section, which dissipate the sun's heat. In experiments where the hairs have been removed, the hairless ants reached temperatures 5–10°C higher than the hairy ones—a potentially fatal disadvantage in an environment in which it's essential to keep your cool.

Saiga antelope

Saiga tatarica

How many girlfriends should a respectable antelope bachelor have? For harem-breeding ungulates, such as the saiga antelope, *Saiga tatarica*, this may be a question not merely of masculine boasting rights, but of species survival.

Saigas are nomadic antelopes the size of a German shepherd dog; their large, thousand-strong herds migrate across the semi-arid deserts of Central Asia, mainly Kazakhstan, Mongolia, and Russia. Their most prominent feature is a bulbous, trunk-like nose, doubling as a temperature regulating system for unforgiving prairie weather. In winter, before the frigid steppe air enters the body, the schnoz heats it up; by contrast, it does the cooling in the summer.

Steppe life is harsh, and to maximize the offspring's chances of survival, the antelopes come together for mass calving; within a week, tens or even hundreds of thousands congregate to give birth. This approach helps in two ways. Flooding an area with such huge numbers increases the survival odds of individual animals, who often fall prey to wolves. Meanwhile, a compact birthing period allows the calves to make the most of the short availability of luscious grass.

Young saigas are precocial (see European rabbit, page 34); they are proportionally the biggest calves of all wild ungulates, and are able to outrun a predator when they are just a few days old. Still, producing such well-developed calves puts a lot of strain on their moms, making them particularly vulnerable to diseases during this time. In 2015, 200,000 animals, over half of the global saiga total, succumbed to the usually benign bacterium *Pasteurella multocida.* Other mass mortalities, due to peste des petits ruminants virus and other diseases, are not uncommon, and saiga populations tend to fluctuate over decades. Thankfully, while their mortality can be extremely high, it is balanced out by high fecundity, with most females birthing twins each year.

Saigas practice polygyny: a male mates with many females, but the female mates with only one male. While the females bear the brunt of childcare, the males compete for mates—and are therefore under the selection pressure to be bigger. Bigger body means better chances of fighting off competitors, accruing a larger harem, and achieving greater reproductive success. Consequently, like most other polygynous species, saigas are sexually dimorphic: the males are bulkier (about 1.5 times the size of the females) and sport a set of thick, slightly translucent horns. Unfortunately, these very horns get them into trouble.

Unluckily for saigas, their horns are highly valued in traditional Chinese medicine; they are used as a replacement for rhino horn, believed to have the same healing properties. After the collapse of the Soviet Union—which did a decent job of protecting saigas and regulating commercial hunting—the trade with China and other countries opened up, and poaching and hunting nearly led to the species' extinction in the early 2000s. It is mainly the males who are being targeted, as they are harvested to meet the high demand for horn.

Under normal conditions, there would be one saiga male to four or five females; however, with increased sex-biased poaching, this ratio changes. One would assume that a lower male-to-female ratio would simply mean fewer very busy and pleased males—but there is more to it than that. Conservation scientist E. J. Milner-Gulland discovered that while having only 5 percent of males in a population still allows saigas to flourish, a proportion somewhere below 2.5 percent leads to a population collapse. When the ratio got as low as one male to 106 females (0.9 percent of males in the population), the females were less fecund: there simply weren't enough males to fertilize them. It reached a point where the courtship behavior was reversed—the does were actually competing for access to bucks. The older, more dominant females fended off the younger ones, resulting in their failure to conceive.

Between mass mortality-inducing diseases, and male-selective poaching, it is not surprising that saigas are critically endangered. Being a horny horned antelope is a dangerous business.

Slave-making ant

Temnothorax americanus

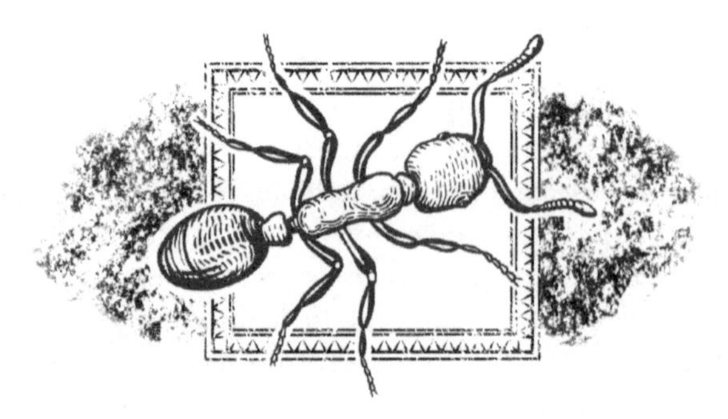

A New England woodland might seem like the ideal place for a tranquil stroll, but, unbeknownst to Sunday amblers, just a few square meters of forest floor suffice to witness battles, ensnarement, societal downfall and the evils of slavery. All on a micro-scale, because both the assaulters and the victims are ants.

While ants are abundant both in sheer weight (estimated at up to 20 percent of the mass of all animals that live on land) and in the number of species (some 14,000), only about fifty or so types resort to enslaving others. Like Vikings, these ants pillage nearby colonies, in a process known as dulosis, from the Greek *dulos*, "slave." They are biologically adapted for one task: to subdue colonies of other species. As social parasites, they rely on other ants to forage, care for their young and defend their nest. Some species, such as the Amazon ants (*Polyergus rufescens*), are not able to even feed without the help of slaves, and will starve in the presence of food if it is not delivered directly to them. Plus, their dagger-like mandibles, while great for terrorising other ants, are useless when it comes to looking after babies.

While the Amazon ants perform extensive raiding operations, involving thousands of warriors, another ant species, the 2–3mm-long *Temnothorax americanus*, plunders on a much smaller scale. Their victims belong to one of three related species, *T. longispinosus*, *T. curvispinosus* and *T. ambiguus*, living in small colonies that fit inside a single hollow acorn or twig—and even a modest raid can have serious consequences for their survival.

A dulotic colony is started by a newly mated queen, who invades the hosts and evicts or kills the local queen and adult workers, sparing just the pupae. She then waits for the pupae to emerge, and, once they do, they work for her: gathering food, protecting her and looking after her young. The slave-maker queen will only lay a couple of eggs in the colony's first year (the number subsequently increases to about ten), and the daughter workers' sole task—and skill—is to recruit more slaves.

These pirate daughters will embark on expeditions scouting out new colonies to pillage. After localizing the perfect target, they either proceed to attack solo, or return to their nest to assemble a raiding party consisting of conspecifics (members of the same species) and enslaved workers. Timing the invasions precisely to coincide with the presence of pupae in the nests of the victims, the slave-maker troops enter the colonies, slay or displace the resident workers and queens, and abduct the pupae and larger larvae. The youngsters are then taken back to the pilferers' nest to join the ranks of slave workers. The entire colony of *T. americanus* is not too crowded: a queen, a few workers and a few dozen slaves—still, it needs regular replenishing.

The pillaging behavior comes at a cost: the invaded ants will try to protect themselves, and roughly a fifth of the raiders will perish during the siege (only 7 percent of the slave-maker queens are able to successfully establish a colony in the first place). Not only do the attacked insects put up a fight; it turns out that the slaves have also figured out how to retaliate after they've been integrated into the colony.

The ant rebels do so by sabotaging their captors' reproductive powers. Even though they will diligently look after slave-making larvae, once their charges reach the pupal stage, the caring attitude is gone. In fact, over two-thirds of the plunderers' pupae will die off, either killed deliberately (quite literally torn apart by the slave workers) or due to neglect. Young queens face a particularly high risk, with over 80 percent mortality. The males are spared, as they do not take part in the slave raids. While this revolt might not help the enslaved ants themselves, it does protect their sisters in other colonies by reducing the likelihood of future raids. It is full-scale evolutionary warfare—all taking place in the undergrowth beneath your feet.

Slow lorises

Nycticebus spp.

If the Owl and the Pussy-Cat from Edward Lear's poem had a lovechild, it would perhaps resemble the slow loris, with its small, furry, tree-climbing body, a round head with huge eyes, and a nocturnal lifestyle—like its parents. Alas, the tale of the slow loris is not like a children's poem at all; on the contrary, it is a horror story, driven by human greed, egoism, and foolishness.

Slow lorises are a group of eight primate species in the genus *Nycticebus*, inhabiting forests from Bangladesh to Indonesia. Their highly mobile joints, flexible backbones and grasping hands and feet make them supremely well adapted to their arboreal existence. These staunch tree-dwellers feed on nectar, sap, fruit, and invertebrates, providing agroforestry farmers with both pollinating and pesticide services. Unfortunately, even though they are able to live on the forest edge, their populations suffer from deforestation: as they cannot leap, they are reliant on trees, vines, and lianas to move from one habitat to another.

Lorises feature prominently in folk beliefs and traditional medicine in Southeast Asia. The animals or their parts are believed to heal over a hundred illnesses, from postpartum ailments and stomach aches to broken bones and sexually transmitted diseases. They are widely sold at marketplaces—dried, turned into tablets and ointments, or even roasted alive to increase their medicinal potency.

Apart from their alleged medicinal value, one more factor seals the fate of the lorises: their undeniable cuteness. These doe-eyed, sluggish primates with a worried look on their faces are frequently used as tourist photo props across the region; they are also exported to Europe, Japan, and Russia as exotic pets. This is despite an obvious unsuitability for the job, since they are the only primates with venom. Known as "the shy ones" in Borneo, because they cover their faces with their hands when threatened, lorises can nonetheless deliver an extremely painful, venomous bite. Their "shy" defensive posture allows them to combine saliva with a secretion from the brachial gland located near the elbow; this cocktail produces a toxin strong enough to induce tissue necrosis, anaphylactic shock and even death.

As the animals used in the tourism industry and pet trade are harvested from the wild (they don't reproduce readily in captivity), they need to be made "safe" for handling. Their sharp teeth are therefore ripped out or cut with nail clippers, causing pain, blood loss and often leading to death. Captive lorises are fed inadequate diets and cannot express their natural climbing behaviors. They are also usually handled during the day, causing further suffering: bright light hurts their sensitive large eyes, and the sun burns their retinas. Even if they are rescued, their permanent injuries prevent their release back into the wild.

In a world full of domesticated kittens, puppies, and rabbits, taking endangered animals from their habitats purely for amusement seems unfathomably pointless and selfish—but the demand for these primates is often driven by their presence on social media. Despite the fact that capturing and trading them is illegal, videos of captive lorises—tickled, dressed, fed popcorn—tend to fuel the appetite for them. Consequently, primate researcher Anna Nekaris proposed that YouTube should tag such videos of slow lorises with "animal cruelty" or "conservation threat" labels, or better yet, remove them altogether. The 2015 awareness campaign "Tickling is Torture" resulted in a brief change of attitudes toward captive lorises, but unfortunately did not seem to have a long-lasting effect.

In the current ubiquitous selfie culture, the desire to have an Instagrammable pet or at least a photo with a loveable primate outweighs the price paid by the animals involved in the process. Sadly, even the seemingly innocuous sharing of captive wildlife visuals on social media fuels the industry.

In Lear's poem, the Owl and the Pussy-Cat are last seen "hand in hand, on the edge of the sand," dancing by the light of the moon. It's an idyllic image that is tragically a long way from the fate of many of their photogenic offspring.

Southern grasshopper mouse

Onychomys torridus

This is the story of a Wild West outlaw who roams the deserts and prairies of Mexico and the American Southwest. Above all rules, feared by everyone, fearing no one, a true desperado—except it's the length of a pencil. It's a mouse.

The southern grasshopper mouse, *Onychomys torridus*, deserves to be called the baddest mouse in the West; heck, maybe even the world. Unlike the house mouse and most other rodents, it's almost exclusively carnivorous. It will attack, kill, and eat anything that might cross its path: scorpions, beetles, grasshoppers, and even other mice. Grasshopper mice are incredibly skilled and well-equipped assassins: their jaws deliver a particularly forceful bite, their fingernails are more like talons (*Onychomys* means "clawed mouse"), and they use a range of techniques for hunting different victims. Fast-moving grasshoppers are executed with a quick chomp to the head. Stink beetles that spray a noxious substance are held with the abdomen against the ground, to block the spread of the defensive secretion. Rodents are killed with a swift bite to the base of the skull, which severs the spinal cord. Yet the most interesting fight is that between a grasshopper mouse and a scorpion.

Scorpions defend themselves by delivering a painful sting, used both as a deterrent and a means of buying time to escape. Bark scorpions, which feature on the grasshopper mouse menu, not only inflict intense pain, but are able to kill a human child with their venom. However, while people stung by this scorpion have compared the sensation to being branded with a hot iron, and afflicted house mice will spend a long time licking the throbbing wounds—the grasshopper mouse will groom the sting injury for just a few seconds before pressing on with the attack. They are not put off their meal even when stung in the face, multiple times. In fact, these berserkers evolved a twisted way to harness the scorpion's venom to their advantage: they use the toxins it contains to block pain transmission, de facto employing one of the world's most painful stings as a painkiller.

Like all true renegades, grasshopper mice are nocturnal; to make their presence known, adult rodents howl in the moonlight. They find a prominent, elevated area, stand on their hind legs, prop themselves on their tails and emit long, high-pitched calls with their heads high and their mouths wide open. The howls can be heard by humans—and no doubt other mice—up to 100m away; larger individuals have deeper voices. Grasshopper mice have been observed howling in this way prior to a hunt—perhaps making the call a battle cry; alternatively, because most callers were reproductively active males, the "wolf howl" could be a night-time wolf-whistle, a long-distance booty call. The species is not sociable—they will find a mate and breed, but apart from that they are widely spread out, which suggests that their night howling could also be done to mark territories.

Family groups, consisting of a pair and their babies, are tightly knit, and both parents look after the young. Females tend to aggressively exclude their smaller partners from the nest for the first three days after birth, but keen fathers look after their offspring once they are allowed back in, grooming them, huddling over them, and defending them. Perhaps counterintuitively, young grasshopper mice get their aggressive behavior from their doting dads, as single moms produce more docile offspring. Likewise, animals fostered by white-footed mice, a meeker species, were found to be much less belligerent.

Reckless, violent, and ruthless—all this gangster-mouse is missing is a Colt (and an opposable thumb to hold it).

Tarantulas

family Theraphosidae

What a pleasant family leaving the house—a few vivacious pets, followed by the children and then, finally, the mom. A delightful picture. It might come as a bit of a surprise, therefore, to realize that this idyllic scene is a party of tarantula spiders.

Tarantulas are generally large, hairy spiders, belonging to a thousand or so species in the family Theraphosidae. Their name is a bit of a mistake—the original "tarantulas" (*Lycosa tarantula*) are a completely unrelated species: the 2–3cm-long wolf spiders from the Italian town of Taranto. Yet, over time, any sizeable and scary spider became a "tarantula," and eventually the term stuck to denote the theraphosids, who are most certainly much bigger and hairier than the wolf spiders.

Tarantulas hold the record for the heaviest and longest-bodied spiders. The Goliath birdeater, *Theraphosa blondi*, has a body length of 13cm, weighs 175g (as much as a week-old kitten) and its legs can span 30cm. Tarantulas also have the sweetest, furry, cat-like paws, with retractable claws used for climbing.

There are other similarities with kitties. Like cats, tarantulas enjoy warm places—they are found all around the world, except in northern parts of Europe, Asia, and North America. Like cats, they can deliver a solid bite: their fangs can reach just under 4cm in length. And also like cats, larger tarantulas can prey on rodents, bats, reptiles, and birds—though they prefer amphibians, arthropods, and worms. However, in contrast to felines, tarantulas' eyesight is terrible, and they prefer to hunt using touch. These spiders are ambush predators: they will sit and wait, and, upon detecting the vibrations of passing animals, pounce and kill. They sometimes have trouble identifying smaller prey items using this hunting method, which explains why they tend to go for larger foodstuff.

Despite their imposing size and ferocious appearance, tarantulas often fall prey to other animals, particularly as they are very protein-rich morsels, with 63 percent protein content (over three times that of vertebrates). They are the staple of some snakes, although mammals such as the coati, or birds such as the domestic chicken, are also partial to a bit of tarantula. To protect themselves, the spiders use both ends of their body. At the front, they have two impressive fangs, loaded with venom (which, while not deadly to humans, can cause pain and discomfort). At the back, on their abdomen, New World tarantulas sport a killer hairstyle—urticating hairs: fine, barbed bristles, which, when flicked in the direction of the attacker, irritate the eyes, nose, or airways. The hairs can lead to permanent eye damage, and even be lethal to small animals.

Still, there are some creatures that live in perfect harmony with the tarantulas, acting a bit like favorite house pets. The ferocious spiders rather enjoy sharing their home with little, narrow-mouthed frogs from the Microhylidae family. Although tarantulas are generally keen on eating frogs, these particular pocket-sized amphibians (most around 1–2cm long) live safely in their burrows. The entire frog–spider patchwork family ventures out peacefully in front of the house for foraging, and returns inside to rest. Even when a spider accidentally bumps into one, it gives the frog a bit of a feel to identify it, and then releases it; the identification is probably done via the skin secretions of the frogs. When faced with danger, the mini-amphibians may run and hide underneath their more imposing housemates, who protect them from predators such as snakes. These tarantula–narrow-mouthed frog associations have been reported across many species around the world; it's suggested that the frogs find shelter and an excellent microhabitat in the tarantulas' abode, while the spiders benefit from the amphibians' ability to eat small pests, such as ants and fly larvae, which could pose a threat to eggs and young spiders. Some spiders keep as many as twenty-two pet amphibians in their burrows, living in cramped but happy harmony.

Tetradonematid nematode

Myrmeconema neotropicum

Ah, ants, the cool kids of the invertebrate world . . . Arthropods want to be them (see Giant prickly stick insect, page 40, and Jumping spider, page 44), parasites want to be in them. Even though one might assume that with their meticulous housekeeping rules ants are impossible to infect, the seemingly impenetrable fortress has been stormed by a sizeable group of creatures: the nematodes.

Nematodes, also known as roundworms, are generally small, thin animals, measuring less than 2mm in length; despite this puny size, they do have digestive, nervous, and reproductive systems. The phylum Nematoda is one of the most abundant taxonomic groups, and may account for even 80 percent of all individual animals on Earth. While around 25,000 species have been described, nematode diversity is estimated at a whopping 10 million.

Roundworms lead a variety of lifestyles, and can be free-living or parasitic. Parasitic nematodes with complex life cycles, involving more than one host, have one main aim in life: to successfully transition from one animal lodging to another. This usually means manipulating the intermediate host to be as edible for the final host as possible. Does it have defenses? The roundworm will deactivate them. Is it normally fast and difficult to catch? The nematode will make it sluggish and disable its escape responses. Or change its colors to be more conspicuous—you get the picture. Still, so far only one parasitic nematode, *Myrmeconema neotropicum*, has had the truly David Copperfield-like capability of turning an ant into a . . . fruit.

In 2005, a research team based in Panama observed a number of *Cephalotes atratus* worker ants with prominent red behinds, in otherwise black colonies. At first, the researchers assumed it was a different species—but upon examining the ants in detail, it turned out that their gasters, or abdomens, were full of eggs, each containing a small nematode. Their story begins when birds infected with the parasites poop out nematode eggs. The eggs are collected by ants and fed to their larvae, and the juvenile roundworms develop in ant pupae. Young adult ants will contain mainly mating roundworms (the females are 1 mm in length, the males are slightly smaller), and, after the males die, the females with mature eggs are the ones left in the ants' gasters. To complete the cycle, the nematodes need to be eaten by frugivorous birds, which is why they turn ant bums into convincing-looking berries.

Infected ants are about 40 percent heavier than non-parasitised ones, and they are also clumsier, slower, and much less aggressive. They no longer bite, and their alarm pheromone production is blocked. Most importantly, the exoskeletons of the ant bottoms become thinner and more translucent, which, combined with the yellowish roundworm eggs inside, creates a bright red color. The now docile ants resemble a delicious berry—moreover, they walk stiffly and erectly, presenting their nematode-ripened gasters to any animal interested in a fruity meal. Finally, again due to infection, the junction between the bum and the rest of the ant becomes 93 percent weaker than in healthy individuals. Consequently, a hungry bird can easily pluck the berry bottom off while the rest of the ant is still attached to its substrate.

It seems that *M. neotropicum* infect the host ants across their entire habitat, namely, Central and South America. Additionally, this type of host–parasite interaction is nothing new, having been around for some 20–30 million years. The earliest evidence lies in Dominican amber, containing ants surrounded by nematode eggs, with abdomens probably pierced by birds. Roundworms have clearly taken their time to perfect the berry-bottom technique.

Texas horned lizard

Phrynosoma cornutum

When it comes to staying safe, a belt-and-braces approach is the best. But the Texas horned lizard's approach is more of a belt-braces-armor-artillery one. It's a reptile that does not take chances.

Native to the deserts and semi-arid regions of the southern United States, Texas horned lizards, *Phrynosoma cornutum*, are also referred to as "horntoads" (their scientific name translates as "horned toad-bodies"). They do indeed resemble toads: they have flat, circular figures and a disapproving look on their wide mouths. But, unlike toads, they are spiky.

With a body length of around 7cm, the prickly reptiles are a tempting snack for many a desert predator, including snakes, birds, coyotes, or even grasshopper mice (see page 70). Yet, despite their size and a reluctance to run when approached by a large animal, horntoads certainly do not make an easy meal—they are veritable security experts.

Their first line of defense is camouflage. Not only are they the color of the dry habitats they live in—reddish, yellowish, gray—but the bumpy skin helps them blend into their surroundings, as does their preference for a sedentary lifestyle. If their cover is blown, horntoads make themselves Big and Scary (as far as that's possible for an animal that could comfortably fit on a saucer). They inflate themselves to twice their original size: the armored pancake turns into a spiky balloon, hoping that the attacker will reassess its swallowing capabilities.

If the size change does not put the predator off, the horns and spines might. While the spines are modified scales, the horns are actual, bony horns—protrusions from the lizard's skull. When facing the inevitable, the horntoad will bow its head, exposing the horns, and make the carnivore's meal as hellish as possible. It works—reports have shown adult roadrunners regurgitating horned lizards, while at least one overconfident young bird choked to death as the lizard's horns pierced its neck from the inside. Similarly, a juvenile rattlesnake has been reported to meet its end after the spines of a horntoad gored its body wall from within.

For larger carnivores, in particular canids, the lizards will use an extra-special deterrent: autohaemorrhaging, or spraying blood from their eyes. The reptiles increase the blood pressure around their head to the point of rupturing blood vessels, resulting in squirts of blood being shot from the ocular sinuses at a distance of about a meter. The blood itself is unpalatable to canids, so a fox, coyote, or dog will think twice before tucking in.

The distasteful compounds within the blood probably come from the horntoads' diet. The lizards' main food source is the highly toxic harvester ant, *Pogonomyrmex maricopa*, topped up with beetles and termites. Horned lizards' plasma can detoxify the harvester ants' venom, but the blood ends up foul-tasting for dogs. However, it is apparently fine for humans, based on (numerous) published accounts. Unfortunately, due to excessive use of pesticides and the spread of highly aggressive, invasive fire ants from South America, the lizards' food source is in decline.

Horntoads wash the ants down with water—and even though this is a scarce resource in their habitat, they have figured out their own rainwater collection scheme. The lizard arches its back in an umbrella-like shape, and its grooved skin uses a capillary system between the scales to direct water straight to the mouth. Rain or shine, roadrunner or fox—the squat little reptile is prepared for all eventualities.

Velvet worms

phylum Onychophora

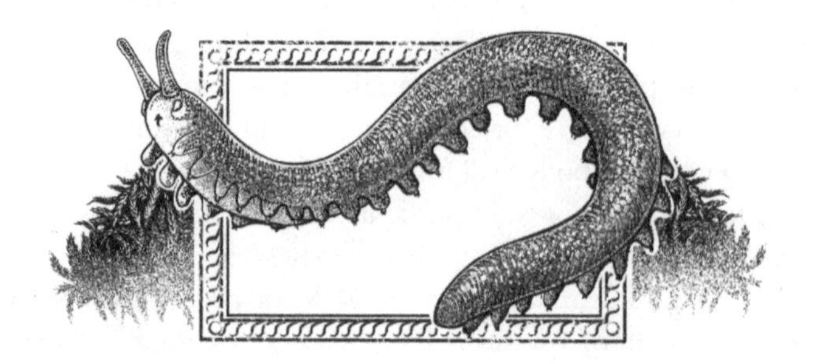

If animals picked their own superpowers, velvet worms chose a pretty good one: shooting out strands of net-like glue to immobilize prey. Unfortunately, not all wishes pan out as intended. Rather than resembling the agile and powerful Spider-Man, velvet worms are a bit more like a dilapidated glue gun.

Velvet worms belong to their own phylum, Onychophora, and they have changed very little since the Early Cambrian period. In terms of phylogeny, they are placed between arthropods and tardigrades (see page 148), though in terms of shape and size they resemble caterpillars with antennae. They have many stubby little legs—between thirteen and forty-three pairs, the numbers differ both between and within species—and a segmented body, but no hard exoskeleton, only a fluid-filled, hydrostatic one. They are also painfully slow creatures, moving at 4cm per minute.

So slow is their pace that when they are hunting—usually for small insects, woodlice, snails, or spiders—their movements are steady enough to allow them not only to creep up on their prey unnoticed, but also to examine it gently using their antennae. If the prey ticks all the boxes, the attack commences: specialized oral papillae shoot a series of squirts of protein-rich glue. Like a glue gun—and unlike Spider-Man—velvet worms can only fire adhesive at short range, at most a few centimeters. Usually, a single jet is enough to immobilize small prey, but larger animals might receive a few extra sticky strands around the legs, while spiders will be shot in the fangs. Once they have squirted the adhesive, velvet worms bond to their prey, literally. They feel up the immobilized critter in search of a soft spot, and use their jaws to inject saliva, killing the prey and starting the external digestion process. While their digestive enzymes stew the main course, velvet worms themselves tidy up the crime scene by eating the dried-up, high-energy glue they produced during the hunt. The ingestion of the victim's flesh takes several hours, and a square meal provides enough food for one to four weeks.

Velvet worms are the only exclusively terrestrial animal phylum, which is a somewhat surprising life choice given that they are prone to desiccation and require habitats with high humidity. They may be solitary or gregarious. One social species, *Euperipatoides rowelli*, lives in female-led groups of up to fifteen individuals, each group rather protective of the rotten log it inhabits. These velvet worms hunt as a pack, and the dominant female has first dibs on the food. After she spends an hour or so dining solo, the remaining females, then the males and the young, are allowed to share the scraps. Subordinates are shown their place through bites, chasing, and kicks.

When it comes to velvet worm reproduction, variety is the spice of life. Some species are oviparous, or egg-laying. Some are viviparous, which means they give birth to live young. Those who want the best of both worlds are ovoviviparous—the eggs are carried internally in the mother's uterus, but sustained from their yolks. There is even a parthenogenetic species. Certain velvet worm males have the most peculiar way of attracting mates—using not exactly their smelly feet, but the next best thing: pheromones emitted via crural glands on their legs. Even more bizarrely, some species deliver sperm packets (spermatophores) by literally giving head—instead of a penis, they have dedicated structures (bumps, dimples, or daggers) on their heads for making the transfer. Perhaps most oddly of all, some velvet worm females do not need to receive sperm in a designated drop-off site. The spermatophore is simply placed on the skin, which triggers a reaction: specialized haematocyte cells of the female break down her cuticle and the envelope surrounding the sperm packet, allowing the release of sperm directly into the body cavity and, subsequently, the ovary. Given how slow velvet worms are, perhaps it's best to keep things straightforward.

Wombats

family Vombatidae

Wombats, the Australian marsupials similar to an oversized, 30kg guinea pig, come in three flavors: common (*Vombatus ursinus*), southern hairy-nosed (*Lasiorhinus latifrons*) and northern hairy-nosed (*Lasiorhinus krefftii*). The latter, critically endangered, species is now restricted to a range of 3 square kilometers within Epping Forest National Park in Queensland. All three wombat species are herbivorous. They live underground and are excellent diggers; to make life easier for their young, the females have rear-facing pouches, which prevent joeys from getting mouthfuls of soil when the moms are burrowing.

Unsurprisingly, the tunnels of such sizeable animals can attract uninvited visitors—but a wombat who is chased into a burrow by a predator, such as a fox or wild dog, wields a pretty efficient weapon: its shield-like rump. A cornered wombat will wedge itself head-first into a tunnel, blocking it so that only its backside is exposed to the adversary. That backside is superbly adapted to protect the internal organs, as it consists of a hard plate made up of four fused bones covered in cartilage, a solid layer of fat, a thick skin and, finally, dense, coarse bristles. While a dingo or a Tasmanian devil could hunt a wombat in an open space, getting through such defensive armor underground is a pain in the arse for both sides. All the while, the wombat does not politely wait to get eaten—not only can it deliver a strong kick or two, but there are accounts of wombats using their bulletproof butts to whack predators against tunnel walls or ceiling, crushing their skulls, or suffocating them in the process. Zookeepers' manuals warn staff against pushing arms between a wombat's bottom and a hard place.

Probably the best-known feature of the wombat is its unique ability to generate cube-shaped droppings. The production process has baffled scientists for decades, particularly because wombats' anuses are not square at all; however, only recently has the shroud of mystery been lifted. It took an international collaboration between researchers from the University of Tasmania and Georgia Institute of Technology in Atlanta to get to the bottom of it. During dissections, the team of scientists observed two things. Firstly, wombats have a relatively long colon—almost 6m, compared to the 1.6m of an average human—which means that the droppings are very dry by the time they reach its end. The prolonged journey through the gut (lasting some 40–80 hours on average) and subsequent desiccation of the poops may allow them to retain their shape better. Secondly, it is now apparent that the feces acquire their geometric shape while still in the colon, and not, as previously suspected, during defecation. Wombat guts are not uniform in elasticity and thickness—they consist of softer and stiffer fragments, as well as thicker and thinner ones. After more dissections, alongside tensile testing and using mathematical models, the study team came to the conclusion that as the poops pass through the stiffer segments with faster contractions, and the softer segments with slower movements, their walls become flattened and their edges sharper. This research, published in the appropriately named journal *Soft Matter*, gained the scientists an Ig Nobel Prize in Physics in 2019. The Ig Nobels—the wackier equivalent of the Nobel Prizes—are awarded for science that "makes people laugh, and then think." While it may seem very much like blue-sky (brown-muck?) research, the wombat poo findings could in fact have applications in manufacturing processes, or clinical pathology.

As for the wombats, the odd shape of their droppings may help them with territorial marking, since edgy poops are less likely to roll away from the designated marking site. The chubby marsupials produce about 80–100 of these little cubes per day, giving the phrase "bricking it" a much more literal meaning.

Wood frog

Rana sylvatica/Lithobates sylvaticus

In a quiet, wintery forest in Alaska, among the still trees, lies an icy pond. Next to it, under a thin blanket of leaves and a thick duvet of snow, there is a brown frog, small enough to fit in the palm of a hand. The frog does not move; it is frozen stiff. Its heart is not beating; its blood is not flowing; its lungs are not taking in air. And yet, when spring comes, the frog will thaw out and hop into the little pond it spent the winter beside—as if nothing had happened.

This totally cool amphibian is the wood frog, *Rana sylvatica* (also known as *Lithobates sylvaticus*), which resides in woodlands from the Midwestern United States, across Canada, all the way to Alaska and the Arctic Circle. It spends seven months per year frozen, with surrounding sub-zero temperatures reaching a low of -22°C in its northern range. If humans attempted this feat, their cells and tissues would almost certainly be burst and destroyed by ice crystals—but the wood frog has it all figured out.

Wood frogs have their own antifreeze system. They use two substances as cryoprotectants, to keep ice formation in check and help preserve the integrity of membranes and macromolecules. The first is urea, which additionally suppresses metabolism; the second is glucose, whose main role is to keep water within cells. Higher glucose content turns the insides of the cells into a non-freezing syrupy solution, which also prevents dehydration. At the same time, less water outside the cells means less ice around them, and therefore less rupture and damage to the tissues. Thanks to these two protective compounds, the frog will happily survive even if two-thirds of the water in its body is frozen.

The amphibians do not reach their ice-cube state abruptly—they have a few weeks' preparation in September or October, when they freeze during the night and thaw during the day. These repeated freezing and thawing cycles probably help to increase the amounts of antifreeze agents in the body. The frog can easily cope with a 250-fold rise in its glucose level without getting diabetes.

Still, the sugar-laden amphibians have a less-than-sweet disposition, particularly as youngsters. Wood frogs mate in early spring; they prefer to breed in ephemeral wetlands, including pools created by beavers. There is a big advantage to picking short-lived water sources: no resident fish to eat the eggs. However, safety from predators comes at a cost, as temporary ponds are prone to drying out, causing the desiccation of eggs and tadpoles. Also, as the pool dries, there is less food left in it, and more contamination from all the tadpoles denizens excreting waste products left, right. and center. For the frog larvae, growing becomes a race—can they develop before the water is gone, or too filthy?

Their parents may give them a bit of a head start by laying eggs as early as possible, which buys some development time. Because wood frogs lay eggs in one big, communal deposition site, the eggs at the center of the mass may be at an advantage: a slightly higher temperature speeds up their growth, and they are better protected from any external danger. Still, to get ahead of the game and ensure survival in an ever-shrinking home, tadpoles may deal with competition in the most ruthless way possible: cannibalism. While they do not munch on other tadpoles under optimal living conditions, when they sense that the pond is getting crowded (via the chemical cues from the ever-increasing amounts of tadpole waste), they will not hesitate to gobble up their conspecifics. In this frog-eat-frog world, it is usually the oldest, most developed tadpoles that will attack the smaller ones or the eggs, in a classic case of cold-blooded murder.

WATER

Amazon river dolphin

Inia geoffrensis

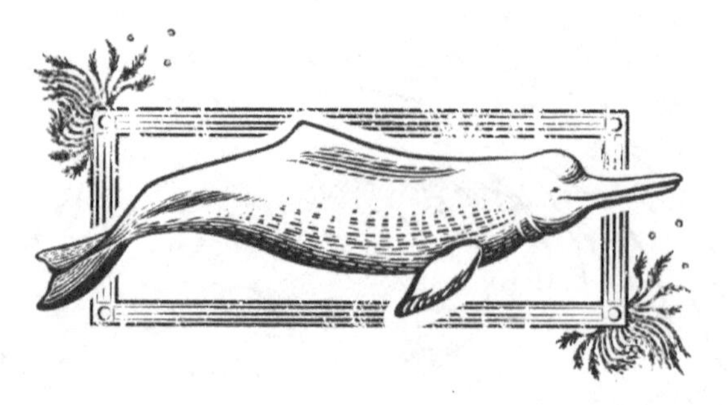

The Amazon river dolphin, or boto, is likely to be very high up on a stereotypical little girl's favorite species list: it's a dolphin *and* it's pink. However, this species has a lot more going for it than its Barbie-doll appeal.

Amazon river dolphins, *Inia geoffrensis*, are found in a range of South American rivers, from Brazil and Ecuador to Venezuela and Bolivia. Since the riverine waters they inhabit tend to be murky, botos use sonar to navigate and to locate their prey—they emit clicks and gauge the time it takes for the sound to bounce off various objects. Their bulbous heads contain a structure called a melon—a large cushion of fat, which acts as a sound lens that focuses the sonar. Because the melon shape can be controlled by muscles, the Amazon river dolphins' sonars are very directional; at the same time the clicks they emit are not as powerful as those of dolphins living in the oceans, probably because their environments are much more cluttered.

The riverine melon-heads are pretty bad-ass when it comes to food: they eat piranhas! They also feed on a variety of other fish, as well as river turtles and crabs—and they use bristles on their snouts to help them detect prey in the mud. The dolphins' eyesight is also good, both in and out of water, despite how tiny their eyes appear. Even though the botos don't swim very fast (usually 2–5km/h) and their bodies look clumsy, they are extraordinarily maneuverable and flexible, with the ability to twist and turn at ease. All these adaptations come in handy when moving through narrower streams, small channels, and rapids; during dry season the dolphins might be confined to deep lakes, but when rivers rise, they venture out into flooded areas following the fish, sometimes swimming among submerged trees.

The species is the largest of all river dolphins—reaching around 2.5m in length, and a weight of 200kg. Males are up to 55 percent larger than the females, which is unusual in cetaceans (the dolphin and whale order), as in most species the females are the bigger sex. Male botos are also pinker, at least to start with. Amazon river dolphins are born gray, and progressively change color—they become pink because of the scarring of their skin. Males, being more aggressive, are at times covered almost entirely in scars, which gives them their rosy complexion.

Before the species' size differentiation became obvious to researchers, it was assumed that Amazon river dolphins were monogamous. However, upon realizing which sex is which, it became apparent that botos have a much more varied love life. There are reports of polygamy, promiscuity, masturbation (both male and female), homosexuality, and even male attempts to penetrate another male's blowhole—though this is probably not a common occurrence, as dolphins need to breathe (through that very blowhole) every 30–90 seconds. Perhaps this is not a behavior that should be mentioned to little girls fond of pink animals.

The diverse sexual behavior aligns with the mythology around the botos. The Amazonians believe the dolphins to be *encantado*—enchanted—shapeshifters that will adopt a human form to lure men and impregnate women. This belief offered the pink dolphins a degree of protection, as few dared to threaten such a powerful animal. However, these days the botos are listed as endangered, primarily due to water pollution and river damming, but also conflict with fishermen. Not only do they get tangled in fishing gear; they are also deliberately killed for bait. In a twist of fate more perverse than their sex lives, the Amazon river dolphins find themselves used as a lure for the fish they themselves used to eat.

Atlantic horseshoe crab

Limulus polyphemus

What has blue blood and has been around for 400 million years? No, it's not a member of the British royal family; it's the horseshoe crab.

Confusingly, the horseshoe crab is not a crab. It isn't even a crustacean. Despite living in the sea, it is related more closely to spiders and scorpions than crabs and lobsters. There are four species of horseshoe crabs—three inhabiting Asia, and one, the Atlantic horseshoe crab, *Limulus polyphemus*, found on the east coast of North America. These marine arthropods are referred to as "living fossils" as they have persisted in an almost unchanged form for over 200 million years—though the earliest fossil records of similar species date as far back as 480 million years, making them older than dinosaurs.

To humans, the horseshoe crab might look like a creature that got the body plan design totally wrong (though they might think the same of us)—they have legs instead of jaws, genitals in their gills, a mouth between the legs, and eyes pretty much everywhere. In Greek mythology, Polyphemus—whom the Atlantic species is named after—was a giant with one eye; yet horseshoe crabs have at least nine. On top of the carapace, there are a pair of compound eyes, a pair of median eyes, plus three rudimentary eyes; on the underside, two ventral eyes placed near the mouth (probably for navigating during swimming, as the animals use backstroke in the water). If that wasn't enough, there are also light-sensing organs along the length of their tails.

Atlantic horseshoe crab eyes aren't any old eyes—they are the eyes that advance science. In 1967, Haldan Keffer Hartline received a Nobel Prize for his research on the neurophysiology of vision; his studies concentrated on the eyes of these marine invertebrates—particularly their large photoreceptors. Still, this is not the species' only input into biomedical research—they are valued for the unique properties of their blue blood.

The oxygen-binding substance in horseshoe crab blood is not haemoglobin, but the copper-containing haemocyanin, which, when oxygenated, turns blue. From a pharmaceutical perspective, however, a more useful feature of the regal blood is its ability to detect endotoxins. Endotoxins are molecules that sit on the outer membranes of bacteria, and they trigger the immediate reaction of amebocytes—the horseshoe crabs' blood cells. Amebocytes, therefore, became the go-to method for commercial endotoxin detection, and have been used in quality assurance for drugs, vaccines, and implants, as well as environmental factors.

To obtain such a vital substance, Atlantic horseshoe crabs—some half a million annually—are harvested from the ocean, taken to labs and bled. Each crab donates around 30 percent of its blood before being released back into the wild. While this may sound like a small sacrifice for a huge biomedical benefit, mortality rates immediately following the harvests range between 10 and 30 percent, a figure that does not include long-term monitoring. The animals suffer from handling stress, injury, overbleeding, and hypoxia. What's more, the blood harvests reduce the females' reproductive ability—which is particularly worrying as collections often take place during the spawning season. To make matters worse, horseshoe crabs are also used by the fishing industry as bait for whelk and eel.

Conservationists are increasingly concerned about the fate of the species, as its reduction has knock-on effects for other animals. Migrating shorebirds, such as the red knot, feed on protein-rich horseshoe crab eggs as a pit stop. Larvae and juveniles are eaten by shrimps, fish, and crabs, while adults fall prey to loggerhead turtles.

Regulators are faced with decisions on how to manage the species in a way that is advantageous for conservation, as well as for biomedical and fishery interests. While some—albeit less efficient—alternative endotoxin tests have been developed, the concern is that curbing the demand from the pharmaceutical industry might disincentivise protective measures currently restricting the fishing industry, leaving the *Limulus* in a legal limbo.

Bluestreak cleaner wrasse

Labroides dimidiatus

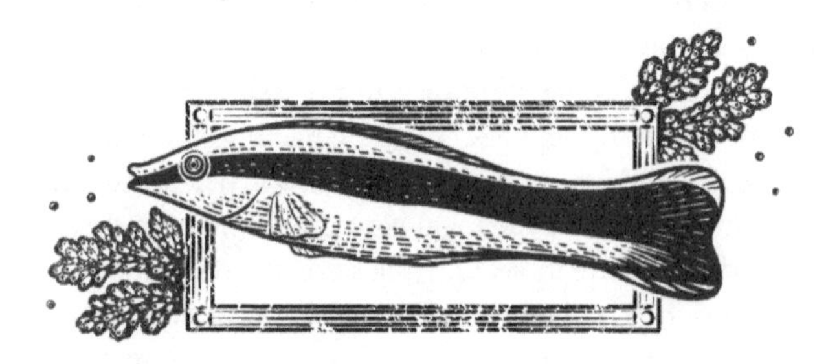

In complex, multidimensional underwater societies, there is a caste of very diligent, hardworking service providers; the salt of the sea, if you will—the cleaner fish. While a number of species engage in cleaning duties, the best studied of these is the bluestreak cleaner wrasse, *Labroides dimidiatus*, found in coral reefs across the Pacific and Indian oceans. These small—about 10cm-long—fish are based at "cleaning stations," attended by larger client fish who swim over to be rid of skin parasites or dead tissue, which the cleaners eat. They don't just service fish—when a sea turtle, octopus, lobster, or seabird comes along, they are equally happy to oblige.

Cleaner wrasses are easy to spot: they have a black stripe along their bodies, against a contrasting blue or white background. They also advertise their availability via a little dance, spreading out their tails and waving their rears. Thus greeted, clients will approach the cleaning stations slowly and calmly, and position themselves in a way that allows the cleaners free access to their fins, gills, mouth, and other body parts. The cleaning business is incredibly successful, with a single cleaner having more than 2,000 interactions a day—so it comes as no surprise that some fish species mimic the looks and behavior of the wrasses. False cleaner fish (*Aspidontus taeniatus*) emulate their bright colors and characteristic black stripe; they will also do the "welcome to our cleaning station" dance. However, false cleaner fish are fin-eaters, and will take advantage of the splayed-out posture of the client to bite their bodies instead of cleaning them. Another aggressive mimic, the bluestriped fangblenny, *Plagiotremus rhinorhynchos*, actually drugs its clients with opioid-containing venom, so they don't feel the bites and are too disorientated to pursue the offender.

One might draw the conclusion that wrasses are kind and helpful, and the mimics nasty and exploitative—but it's not that simple. While cleaner wrasses do mainly feed on parasites, what they really, really love is the delicious, nutritious scales and mucus of other fish. Since clients do not like being nibbled—they will chase the nibbler or avoid visiting the abusive cleaning station—the wrasses face a dilemma: to eat tasty mucus morsels and risk the business going bust, or settle for the suboptimal parasite meal and maintain a constant stream of clients.

It is very apparent when a cleaner is cheating the client—its painful bite elicits a visible jolt. As cleaning is done in the presence of other fish, such jolts can signal: "Watch out, the clientele gets bitten in this joint!," gaining the bitey cleaner wrasse a bad rep and losing it trade. Consequently, cleaners doing their job when other fish are watching tend to be on their best behavior, to encourage future patrons; instead, they bite more when nobody's looking. Additionally, biters who cannot resist temptation, but don't want to lose customers, will offer back massages with their fins to make up for their moment of weakness (and show off the provision of extra services to onlookers).

Visiting clients, who have access to other cleaning stations, get preferential treatment over residents, who don't have a choice. Visiting clients represent an ephemeral food source—and cleaner wrasses quickly learn to take advantage of something before it disappears. In fact, when pitted against primates such as capuchins, chimpanzees, and orangutans—the brainiacs of the animal kingdom—in a task which required maximizing food intake (that is, eating from a temporary food source first, and only then from a permanent one, thus receiving an additional delayed reward), the wrasses learned very quickly, while most primates failed to perform above chance. For fun, Redouan Bshary, the project's lead investigator, set up a similar "foraging test" for his four-year-old daughter, with chocolate M&Ms on temporary and permanent plates—alas, in a series of 100 trials, she never learned to prioritize the temporary plate.

Bobbit worm

Eunice aphroditois

Naming an animal *Eunice aphroditois*, after Eunice, a sea nymph, and Aphrodite, the Greek goddess of love and beauty, certainly puts a lot of pressure on it, appearance-wise. But the bearer of this dainty name doesn't yield to such pressure—it looks like an amalgam of aliens from all horror movies in Hollywood history. It is a humongous, segmented, carnivorous worm. The only redeeming feature that might possibly explain the Aphrodite in the moniker is a very pretty iridescent sheen on the worm's body—beautiful, if you overlook its otherwise monstrous features.

The animal's common name, bobbit worm, is more appropriate for its predatory nature. It memorialises the John and Lorena Bobbitt case, which made international headlines in the 1990s, when Lorena cut off her abusive husband's penis with a carving knife while he was sleeping. Bobbit worms don't chop off each other's phalluses—perhaps because they lack any external genitalia, opting to release their gametes directly into water instead—but they do go for a very sudden attack.

Bobbits are bristle worms, or polychaetes, in the phylum Annelida (segmented worms), which means that they are gigantic, battle-ready, marine cousins of the humble earthworm. They are usually about a meter long, but the largest individuals measure three times as much, and are around 2.5cm thick. There is some confusion when it comes to the exact *Eunice* taxonomy, and currently any oversized, nightmarish polychaete worm from warm waters across the Atlantic and Indo-Pacific seems to be called a bobbit.

Also known as sand strikers, these ambushing predators spend their days buried in sediment, in their mucus-lined lairs. The only body part that pokes out is the head—and it is perfectly armed for a surprise assault. Five striped antennae act like trigger spikes on a naval mine: if a fish swims over, either lured by their vermiform (wormlike) movements or due to an unlucky coincidence, and brushes these feelers, the bobbit snatches it and pulls it into the burrow. As it waits, its strong, razor-sharp jaws, wider than the body, are held open like a spring-loaded trap; when the target is acquired, they snap shut so vigorously that prey may be cut in half. When not feeling around for prey, the sand striker can also use a pair of eyes to locate potential victims. At night, the bobbit changes its hunting strategy to a more active one, protruding from the seabed to grab passing fish.

Some fish, such as the Peters' monocle bream (*Scolopsis affinis*), have learned to stand up to the sand striker. When the worm is spotted, several fish will mob it by blowing water in its direction until the bobbit retreats underground. The mobbing behavior serves several purposes: apart from being an immediate menace to the polychaete, it also informs other fish in the area of its presence, sending a clear message: "You're busted! Time to move elsewhere!"

Speaking of moving . . . during their early life stages, bobbit worms are planktonic and travel around readily, eventually stumbling into the nooks and crannies of rocks or corals. Because of that, they sometimes find their way into aquaria, particularly with items collected from the wild. Sand strikers can spend years living in the tank undetected, and the aquarists only become aware of their presence when valuable aquarium inhabitants go missing. You can imagine the opening scenes of a horror movie: in a serene tank, a fish disappears every now and then—nobody knows why, until one night, an enormous, alien-like underwater worm rears its ugly head, and . . . *snap!*

Deep-sea anglerfish

suborder Ceratioidei

We all know couples like this: she's successful, independent and striking, while he's just a bit of a tagalong. And ain't that the truth for anglerfish.

Deep-sea anglerfish are a group of 168 species in the suborder Ceratioidei, inhabiting the world's oceans at depths below 300m. They include fish with the strangest looks—and names to match, such as footballfish, needlebeards, or warty, whipnose, and prickly seadevils. Anglerfish consist of a gargantuan open mouth with menacingly pointed teeth, a drab brownish-gray body, and a dangly fishing rod growing out of the head. That fishing rod (a modified dorsal fin spine) contains a mesmerizing, luminescent lure, called esca, used for enticing prey straight into gaping anglerfish jaws. What makes the esca glow is symbiotic bacteria, which provide anglerfish with light in return for shelter.

At first glance, a light lure might seem a strange choice—surely fish are not tempted by glowsticks? After all, who shines light underwater? Turns out that the real question is who doesn't?, since an estimated three-quarters of ocean-dwelling organisms are bioluminescent. The light lure is not only attractive to potential prey, it is also useful for locating mates.

The shining and luring is only done by the big and industrious females; in comparison, the males are tiny and nondescript. In fact, adult males of *Photocorynus spiniceps* are so small that, at 6–10mm, they are considered one of the smallest vertebrates in the world (the females reach over 50mm in length). The most pronounced difference between sexes is found in *Ceratias holboelli*, with the female weighing an incredible half a million times as much as the male.

But finding a mate in dark, oceanic depths is not easy, even when she is illuminated with a bacterial lighting system. Male anglerfish maximize their chances with special adaptations: apart from extra-large eyes, they have huge nostrils for homing in on any enticing pheromones. Predictably, once a male finds The One, he does not let go—his normal teeth are replaced by pincer-like bones used specifically for latching on to a female. The males of some species cling on temporarily, but others go a step further and attach themselves permanently to their partners. At this stage, perhaps she's not so much a "partner" as a host, since the behavior of the male is known as sexual parasitism. In a "what's yours is mine" move, the body of the male fuses with the female's, the dermal tissues connect, the circulatory systems become integrated, and his eyes and nostrils degenerate—making him nutritionally dependent on her, till death do them part. Free-living, bachelor *C. holboelli* males don't even feed on their own as adults, but use up all the energy previously stored in their livers to find a mate; from that point onward, they leech nutrients off their sugar mamas.

Icelandic researcher Bjarni Saemundsson, who described this strange phenomenon in 1922, was convinced that the attached males were baby fish. Actually, it's the opposite—a male does not mature *unless* he becomes attached to a female. If he is not lucky enough to find one within a few months, he dies. The female may be parasitised before she is sexually mature; from her perspective, it is practical to have a pocket-sized male on hand, even if he isn't much company. Whenever she is ready to make babies, he is around—and together they form a strange, hermaphroditic, self-fertilizing chimera.

For two separate organisms to become one, in body (and presumably in spirit), they need to not kill each other, immunologically speaking. That is no easy feat, considering that an organ transplant would fail if the recipient's immune system elicited a strong reaction against it. Anglerfish circumvent the problem by pruning down their immune response altogether. In species where mates attach temporarily, the immune reaction is only somewhat weakened, but those with permanent fusion have even lost some responses considered essential for vertebrates. It seems that when a male anglerfish promises he'll be yours forever, he really means it.

Ducks

family Anatidae

Everyone knows what a duck is, right? That is, everyone except for taxonomists. Ducks are part of the family Anatidae, which they share with geese and swans—and that's as far as their classification goes. On their own, ducks don't form a monophyletic group, i.e. they are not composed of descendants of a single common ancestor; instead, they are a "form taxon:" a group classified based on their morphology and behavior. Basically, most taxonomists go with the good ol' abductive reasoning test of: "If it looks like a duck, swims like a duck, and quacks like a duck, then it probably *is* a duck" to classify what is and what isn't one.

To make matters worse, ducks are prone to a bit of inter-species intermingling. Mallards (*Anas platyrhynchos*) mate with over forty other species, including the endangered Hawaiian ducks; ruddy ducks take a shine to the endangered white-tailed ducks; overall, more than 400 interspecific hybrids have been recorded in waterfowl. This is a serious conservation concern, especially on islands where the populations of native ducks are small. Hybridisation occurs so frequently because of similar genetic make-up and behavior between species, as well as sharing the same habitat. However, one other thing might be a contributing factor: ducks belong to the 3 percent of bird species in possession of an external penis.

Most birds are rather chaste: because they lack penises, they reproduce by a "cloacal kiss." Much like naked Ken and Barbie, they simply press their private parts against each other for a few seconds, just long enough for sperm to go from the male to the female cloaca—et voilà, that's how baby birds are made.

Except ducklings—ducklings are the fruit of an evolutionary war of the sexes.

While ducks are generally considered monogamous (at least for a season), in reality they engage in numerous extramarital affairs, leading to fierce sperm competition. The fiercer the competition, the larger the duck penises. The Argentine lake duck, *Oxyura vittata*, boasts the longest phallus of any bird—the record holder's organ was 42.5cm long—which is also the largest penis of any vertebrate, relative to body size. Moreover, it is covered in spines, likely to help remove the predecessor's sperm from the female's reproductive tract. Duck lovemaking is nothing short of explosive—the Muscovy, *Cairina moschata*, everts its 20cm-long, corkscrew-shaped organ into the female's cloaca at 1.6m/s. Because forced copulations are not uncommon in waterfowl, females try to retain some control over who fertilizes them by developing incredibly intricate vaginas, with deceptive nooks, crannies, and dead-ends. These internal labyrinths spiral clockwise, to make it harder for the anti-clockwise-coiling penis to reach its destination—and indicating that vaginal convolution is a result of sexual conflict rather than cooperation.

It is unsurprising that lady ducks are wary of males—drakes have been reported to force themselves on both females and males, alive or dead. A 2001 account of homosexual necrophilia in the mallard reported a copulation attempt lasting seventy-five minutes—and is not the only account of either homosexuality or necrophilia in this species.

With such exciting love lives, some birds don't have much time to look after the resulting offspring. The black-headed duck, *Heteronetta atricapilla*, doesn't even bother to make a nest—like a cuckoo, it lays eggs for other birds to incubate. Uniquely, and luckily for the hosts (coots, gulls, and even birds of prey—the cheek!), the ducklings are precocious and leave the nest within a day of hatching, without harming their step-siblings.

Sometimes foster care can bring unexpected effects. Hand-reared Australian musk ducks, *Biziura lobata*, learned to expand their vocal repertoires beyond the boring "quack." Not only can this species imitate the Pacific black duck and a slamming door, but one male musk duck named Ripper was recorded imitating his keeper, saying "You bloody fool!"—proving that ducks have the foulest mouths of waterfowl.

Flukes

Microphallus spp.

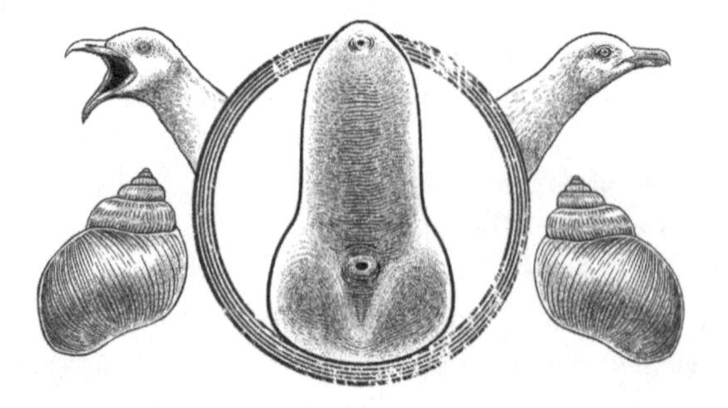

Do not try to look up images of these species—at least not at work! The genus name, *Microphallus* (from the Greek, meaning "tiny penis"), does not leave much to the imagination when it comes to the shape of these critters. The *Microphallus* flukes, like all flukes, are trematodes—parasitic flatworms—distant relations of the better-known tapeworms. They are elongated, flat, and vary in size between 1mm and about 7cm. Their body surface allows direct gas exchange, eliminating the need for a respiratory system. By the same token, flukes consider bums obsolete: food gets ingested, then digested, and finally egested via the opening it came from; the mouth doubles up as an anus.

What flukes lack in the digestive and respiratory departments, they make up for in reproductive strategies. They are hermaphroditic, each animal possessing two testes and one ovary—they can thus mate with a fellow fluke, or fertilize their own eggs. They also can (and usually do) reproduce both sexually and asexually. On top of that, the intricacies of the trematode life history make our own existence look pitifully primitive.

It all starts with an egg. When a mommy and daddy fluke love each other very much (or alternatively, and more likely, when a lonely fluke loves itself just enough), they lay eggs inside the cosiest place they can find—usually the digestive system of an unsuspecting vertebrate. This vertebrate, typically a bird, is called the primary, or definitive, host. Its job, from the fluke's perspective, is to provide a nice environment for adult *Microphalluses* and then poop out their eggs, releasing them into the big, open world. Ideally the big, open world is near water, which is why water-loving birds are often the definitive hosts. There, flukes can find their next victim, the intermediate host; that unsuspecting creature is generally a snail or another mollusc. The eggs can be eaten by the snail, or develop in water to release larval forms which then infect the host and multiply asexually. Flukes with less complex life cycles (e.g. *Microphallus piriformes*) then count on their intermediate host being eaten by the definitive host to allow them to complete the reproduction; flukes with a more complicated lifestyle (e.g. *Microphallus claviformis*) enter a second intermediate host, for instance a small crustacean, before reaching their final, avian destination—where, as adults, they reproduce sexually.

In orchestrating these byzantine scenarios, flukes don't leave things to chance. They encyst in the brains of intermediate hosts and manipulate them in a way that leads them, against their instincts, into the hungry mouths of the final hosts. For instance, fluke larvae plant themselves in the brains of the crustaceans *Gammarus insensibilis*, and steer their underlings from there. They change the crustaceans' response to light, touch, and gravity, which leads to aberrant escape behavior and a higher probability of being eaten. Similarly, infected New Zealand freshwater snails (*Potamopyrgus antipodarum*) are more likely to spend time foraging on top of rocks in the morning, when waterfowl, the flukes' preferred host, can eat them—but less time in the afternoons, when fish (an unsuitable host) are out on the prowl. On the other side of the world, in Scotland, infected periwinkles (the snails, *Littorina saxatilis*, not the flowers) show analogous behavior—they are more likely to move upward, into dangerous gull territory, than uninfected conspecifics. Interestingly, this pattern prevails in snails with mature infections, when the parasites are ready to move on to a new host—periwinkles with immature, or non-transmissible, infections do not show the urge to climb on to exposed rocks.

However, flukes do not restrict themselves to altering the behavior of their hosts. In what can only be described as a "dick move," the *Microphallus* castrates the periwinkle by occupying its gonads. Parasitic castration leads to increased growth of the snail—providing the fluke with greater resources for its own reproduction. For creatures with very basic brains, these trematodes are capable of truly Machiavellian plans.

Gharial

Gavialis gangeticus

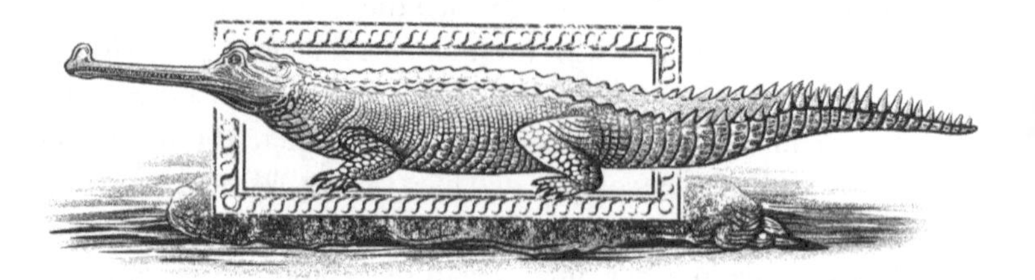

Looking for a nanny? Consider this candidate: caring and delicate, fun and funny-looking (cartoon-like snout, googly eyes, very long body), loves belly slides and water play, nose ideal for booping. Practically perfect in every way; it really is a coin toss between Mary Poppins and the gharial, *Gavialis gangeticus*.

Despite being the second largest crocodilian—with males reaching lengths of over 6.5m—it is much gentler than the record-breaking saltwater crocodile (*Crocodylus porosus*). The gharial, native to the rivers of South Asia, is a pescatarian, and uses its comically skinny snout armed with around 110 teeth to catch fish. Apart from being the toothiest crocodilian, it is also the most thoroughly aquatic, leaving water only to bask and lay eggs. An outstanding swimmer, on land it resorts to sliding around on its stomach while pushing off with its thin limbs.

Unlike crocs or alligators, gharials can be sexed at a glance. The male, upon reaching sexual maturity at the age of eleven years or so, starts growing a bulbous knob at the end of his nose. This growth resembles a traditional Indian earthen pot called a *ghara*, hence the name of the species. The nose protuberance serves to amplify the hisses emitted by the males' nostrils; on a clear day, these snorts can be heard almost a kilometer away. Gharials are polygynous, and males guard a territory occupied by several females.

Gharial moms lay the biggest eggs of all crocodilians, weighing around 160g each, three times as much as chicken eggs. A clutch of up to sixty is sealed in a nest dug in a sandbank, with the hatchlings excavated by their mothers in response to their chirps. The youngsters from one clutch stick together, although offspring from several nests will attend one "creche" of 120 or more baby gharials, looked after by one or more adults. Childminders are both females and very large males (over 5m in length), who are ready to defend the babies from any threats. The little ones communicate with their carers via vocalizing, while the adults will also use visual displays. They are looked after most intensely for the first nine weeks, but big dads tend the young ones even at the age of nine months or so.

It is in the sandy nests that baby gharials become boys or girls. The gharials' sex determination differs from humans': unlike the chromosomal system, where XY is male and XX female, determined by the genetic make-up of the youngster, the sex of crocodilians is determined by the environment—more precisely, temperature. Fertile clutches develop between 29°C and 33.5°C, and eggs incubating at low (<31°C) or high (>33°C) ends of the spectrum will produce mainly females, while those around 32°C will be mainly males. Also, the higher the temperature, the faster the eggs develop. Crocodilians in general show the FMF pattern (males developing at intermediate temperatures), while most turtles follow a more straightforward, MF pattern—low temperature for males, and higher for females.

Temperature-dependent sex determination is all well and good until, of course, a species finds itself short of one sex because of climate change. As if this wasn't enough, gharials, a critically endangered species, are threatened by a whole variety of factors: pollution, habitat loss, hunting for skins and traditional medicine; they are also much hated by the fishing industry. Gharials with pieces of fishing net tangled around the snout can take as long as a year to slowly famish, while some fishers intentionally mutilate them by chopping their snouts off and letting them starve to death.

Thankfully they breed readily in captivity—but more research is needed on gharial caretaking intricacies, to determine when and how the youngsters can be released into the wild. These devoted babysitters need their own protection.

Giant Australian cuttlefish

Sepia apama

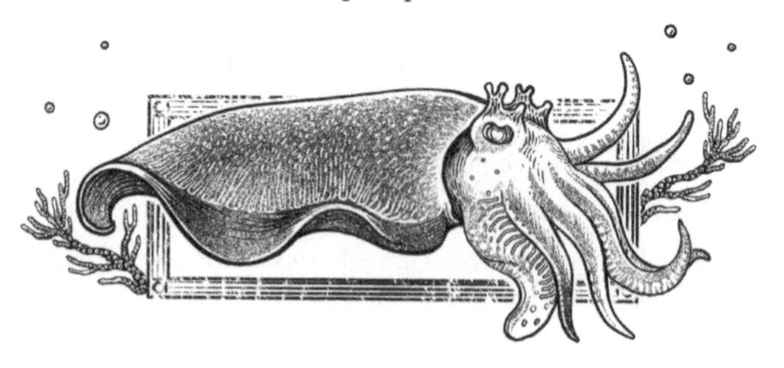

To what lengths will males go to have sex? Impressive ones, at least if they are cuttlefish males.

Take the giant Australian cuttlefish, *Sepia apama*, as an example. Solitary for most of the year, in winter they come together in spawning aggregations. The most spectacular of these is found in Spencer Gulf, South Australia, with over 40,000 cuttlefish occupying a mere 8km stretch of coastline. Upon arrival, cuttlefish—who, like most other cephalopods (i.e. octopuses, squids, and nautiluses), are excellent at swift color changes—will put on their finest displays. Males abandon safe, camouflage coloration and pulsate bright zebra patterns across their skin to attract females—who don the mottled design—and they are ready for action.

Both sexes will mate with multiple partners, multiple times. Cuttlefish sex is a face-to-face or, rather, head-to-head affair. The female will welcome the male with ten open arms, and he starts off by squirting water from his funnel into her mouth to clear out any competing sperm. He then uses his hectocotylus—a specially modified fourth arm—to deposit a sperm package below her beak and break it open in her buccal cavity. The female fertilizes the eggs herself, immediately before laying them, by passing them over the sperm receptacle in the mouth region. As she can take a bit of time to lay, the male guards her to ensure paternity.

This process may already sound convoluted; however, there is an added level of complexity to the cuttlefish love games. While the overall proportion of males and females in the population is roughly equal, the gents spend much more time at the spawning site—which means that the de facto operational sex ratio is roughly four to one, reaching a peak of eleven lads to a lady. Consequently, there is fierce competition for cuttlefish males to find a mate—and they will resort to all sorts of strategies to get her.

The larger ones will challenge a paired male directly, by putting on their manliest displays and starting a head-on confrontation, sometimes successfully. However, the smaller ones don't stand much of a chance of winning such a challenge, and so they resort to stealth—akin to the "sneaky fucker strategy" of the common side-blotched lizards (page 32). They can turn either to open stealth, approaching the female as her partner is fending off the manly males, or to hidden stealth, waiting under a rock to score a date when she's about to lay eggs.

The smallest males have one more option: cross-dressing. To reach a couple undetected, they will put on the feminine mottled skin patterning, hide their sperm-wielding fourth arm and adopt the posture of an egg-laying female. The disguise is so good that it not only fools the other males (the big ones become overly protective of this new arrival, while the smaller ones attempt to mate with him), but it also had scientists scratching their heads during behavioral observations.

Females, meanwhile, can be rather choosy, rejecting over two-thirds of mating attempts. When they are too busy laying eggs, or just not interested in sex, they put up an "engaged" sign—a bold white stripe along the base of the fin. The males who ignore this warning are unceremoniously pushed away and occasionally sprayed with ink.

The tactics of the giant Australian cuttlefish may be elaborate enough, but the mourning cuttlefish (*Sepia plangon*) one-up them. Like a cabaret act featuring a solo performer dressed in either male or female costumes when viewed from different angles, these molluscs can display a male pattern on one side of the body, and a female on the other. A male can thus woo a lady to his right, while simultaneously appeasing her boyfriend on his left. Interestingly, the half-and-half act is only used in the presence of one rival, perhaps because two males might be too difficult to deceive. Truly, cuttlefish take catfishing to a new level.

Giant water bug

Lethocerus deyrollei

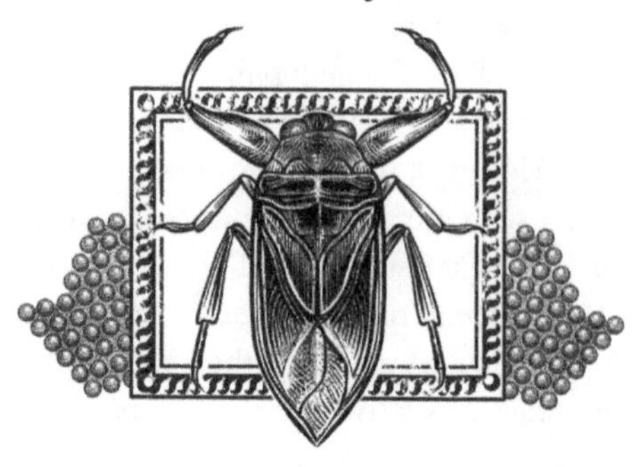

Fish eat insects—that sounds about right. But what about insects that eat fish? Enter the giant water bug, the largest aquatic insect—and one of the most voracious. Giant water bugs, also known as toe-biters, encompass around 170 species in the family Belostomatidae; they are grayish-brown, flattened, and leaf-shaped, which allows them to blend in with their pond habitat while waiting for unsuspecting prey. They come in two varieties, or sub-families: the smaller Belostomatinae, about 2cm long, and the larger Lethocerinae, reaching up to 12cm. While the petite toe-biters feed on insects, crustaceans ,or snails, the menu of the biggest toe-biters features an array of vertebrates, from fish and amphibians to snakes, ducklings, and turtles.

The giant water bugs come wonderfully prepared for their predatory lifestyle. Like all insects, they have six legs; the middle and hind pairs are used for swimming, while the forelegs, which look like the arms of a cartoon Superman flexing his muscles, are adapted for grabbing prey. Like all true bugs, toe-biters practice extra-oral digestion. This means that they pierce the body of their victims with their sharp rostrum (beak), inject digestive enzymes, wait a bit, and then use the same straw-like mouthpart to slurp the fleshy-slushy while the prey is often still alive. Even though their bite does not cause humans lasting harm, it is noted as one of the most painful bites delivered by an insect.

Young giant water bugs learn to hunt big game the hard way—during the breeding season, smaller prey is absent, so the tiny toe-biter nymphs attack tadpoles and fish fry much larger than themselves. Yet before they are ready to hunt on their own, they are looked after by their fathers. Giant water bugs are exemplars of paternal care in arthropods: the role of the mom, by contrast, boils down to laying the eggs; she spends her adult life alternating between feeding and mating. The small Belostomatinae lay eggs directly on the backs of the males, who make sure their backpack nursery is kept moist and aerated by stroking the eggs and swimming up to the surface as needed. The large Lethocerinae, such as *Lethocerus deyrollei* from East Asia, deposit the clutches on nearby vegetation. Dads climb the plants to sprinkle the eggs with water carried on their body surface, shade them from the heat, protect them from predators such as ants (using a chemical defense—they release a smelly substance that acts as a deterrent), and also guard them against females of their own species. *L. deyrollei* females will engage in infanticide: when they are ready to mate, but there are no free males around, they destroy eggs to get rid of competition, and to gain a nurse for their own babies. Brooding males will attempt to protect their clutches by attacking the intruder with their front legs; the aggressive (and bigger) female will respond in the same way. Things may get ugly when both sides resort to nipping and start using their rostra; sometimes males sustain serious injuries. To avoid the awkwardness of unwanted advances, some males spend more time than necessary with their eggs in the vegetation above water, where they remain undetected by the lust-crazed females.

Although, when confronted, the male will try to defend the young for some time, there comes a point where he apparently decides "screw this," abandons the eggs and gives in to the temptation of the uncompromising seductress. Emulating a cheesy romance, the giant water bugs abruptly stop fighting and proceed to copulate. The love drama starts again—albeit with a new clutch of eggs.

Greenland shark

Somniosus microcephalus

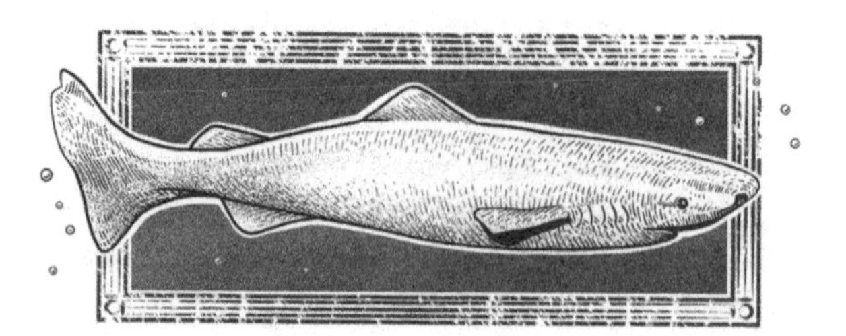

In a broad evolutionary sense, sharks (or their immediate ancestors, dating back around 420 million years) are older than trees (which go back some 385 million years). Meanwhile, present-day individual Greenland sharks have been alive, perhaps not since before trees as such, but certainly since George Washington allegedly chopped down a cherry, and Isaac Newton sat under the apple tree.

Greenland sharks, *Somniosus microcephalus*, from the cold waters of the North Atlantic, are the world's most long-lived vertebrates—potentially reaching an astounding 500 years of age. How does one calculate the life span of such ancient creatures? Conventional aging techniques, such as examining the growth rings on bones, don't work on sharks—their skeletons are made of cartilage. But an international team led by the Danish researcher Julius Nielsen tried radiocarbon dating instead.

In the 1950s and 60s, thermonuclear tests left bomb-produced radiocarbon in the atmosphere, which has since been incorporated into marine food webs. Traces of this radioactive carbon can be found in the eyes of Greenland sharks born around or after these events, because lens tissue proteins don't change throughout their lifetime. The team examined the eyes of sharks of different sizes, from 81cm to over 5m, and estimated the growth rate on the basis of radiocarbon traces. The biggest (and oldest) animal in the study was aged at 392 +/- 120 years. However, being the largest fish species in arctic waters, these slow-growing animals can get even bigger—with record holders reaching 6.4m and weighing just over a ton—indicating that some Greenland sharks out there could be even older.

Nosy scientists are not the only ones looking into Greenland sharks' eyes. They also attract parasitic copepods such as *Ommatokoita elongata*, a type of crustacean that attaches directly to the eyeball. The parasite is unique because of its large size (females with egg sacs measure 4–6cm) and ubiquity—some 98.9 percent of Greenland sharks have them. The copepods infect one or both eyes, anchoring themselves to feed on corneal or conjunctival tissue. They affect the eyesight of the sharks, often blinding them; because of how widespread the parasites are, an uninfected, sharp-eyed shark belongs to a really elite club.

Thankfully, Greenland sharks do not seem to use their sight much when hunting, instead resorting to smelling out prey. This makes sense: they live in low-light environments, from near the surface of ice-covered seas to the dark waters at depths of 1,200m. Based on their stomach contents, Greenland sharks eat fish and cephalopods, as well as mammals such as seals and small whales—but how they hunt is a bit of a mystery. Their blindness isn't the biggest problem: their speed is. Or, rather, their lack thereof. As large, cold-blooded animals, living in very low temperatures, they are constrained to swimming at about 1km/h— the slowest of all fish. Such snail's pace raises the question of how they could possibly hunt the fast-moving arctic seals—one suggestion is that they catch ones that have fallen asleep in the water (which levels the playing field a bit).

As top predators, Greenland sharks' bodies accumulate high concentrations of contaminants, such as PBC or DDT. They also naturally contain high levels of trimethylamine *N*-oxide, a substance thought to improve buoyancy, act as antifreeze for body fluids and protect proteins from deep-water pressures. Add to that high levels of urea, and the result is a toxic meat that can only be eaten after processing. The most infamous way to do so is via months of fermentation, resulting in the Icelandic delicacy *kæstur hákarl*—which, smelling strongly of ammonia, is very much an acquired taste.

As more arctic ice is lost, and more areas become available for fishing, the shark increasingly falls victim to by-catch—and, because the species is difficult to study, it is uncertain how much harvesting the population can take. It is not likely to replenish itself quickly: estimates show that the females do not reach sexual maturity until they are 150.

Hagfish

family Myxinidae

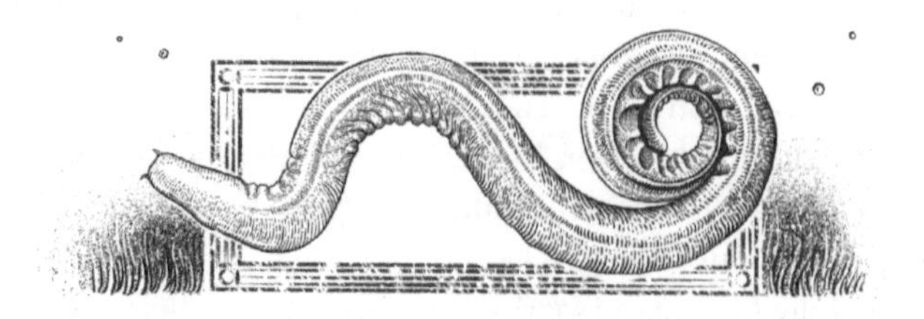

The movie *Jaws* would look very different if the cast members were not people, but hagfish. Imagine the opening scene: a shark approaches its prey, lunges forward for the attack, grabs a writhing body, and . . . spits it out in disgust, choking, spluttering and trying to clear its mouth and gills.

So who are these death-defying creatures? The hagfish (some seventy species in the family Myxinidae) are a type of jawless fish. They are vertebrates who used to have bones, but lost them secondarily over the course of evolution; their boniness is limited to a cartilaginous skull, notochord (a precursor of the vertebra), and fin rays of the tail. They are usually about half a meter long—though some approach three times that—pinkish or grayish, gill-breathing, and four-hearted.

They might lack a backbone, but these primitive fish relatives possess an effective way of repelling sharks, wreckfish, and other large predators. It's slime, also called the hagfish defense gel. Each animal is armed with a battery of slime-producing glands and 90–200 slime pores running down its sides, gaining hagfish the somewhat unkind nickname "snot eels." When disturbed or harassed, they spew copious amounts of goo—structured like no other goo. Hagfish slime, similarly to that of, say, slugs (see page 18), consists of the gel-forming protein mucin; however, it also, uniquely, contains long, thread-like protein filaments. These filaments, about 1–3 micrometers in diameter, are packed tightly into skeins measuring a fraction of a millimeter across. But upon contact with sea water, they unravel to about 15cm, and, together with mucous vesicles, form a tight, fibrous network—ten times the strength of nylon. These slime sheets are not completely impermeable to water, but they certainly slow down its flow, acting like a very fine sieve—and, by doing so, they very efficiently clog the gills of predatory fish. The gel-release takes mere milliseconds to complete, and can be deployed immediately when grabbed by an attacker, before the hagfish sustains any injuries. It is no surprise that any carnivore who was once treated to a mouthful of gag-inducing slime will think twice before attempting to bite a snot eel again.

Hagfish themselves are predominantly scavengers, feeding on fallen carrion and discards from fisheries, or snatching prey from sea cucumbers and starfish. They happily bury into a carcass, entering through the mouth or burrowing directly via the skin. Uniquely among vertebrates, hagfish are able to absorb nutrients directly through their skin, a skill that makes sense for animals that often dine inside their prey. However, they also hunt—using slime, of course. Upon locating a fish, the snot eel corners it and releases enough mucus to slime it to death, by suffocation. Slime, though so universally useful, can also be dangerous to its producer. Hagfish have been known to die when left in their own goo.

To prevent deadly over-sliming, snot eels have developed another unique solution—literally tying themselves up in knots. They are perfectly adapted for it: no constricting vertebrae, no fins that get in the way, baggy skin that fits like a loose stocking and doesn't restrict twisting, while being smooth enough to decrease resistance. The knot on the hagfish body travels toward the head, sloughing off the slime as it does so. This skill is very useful in other circumstances as well—escaping from a tight spot, for instance, or extracting prey from burrows. The knots are also handy when hagfish feed on prey larger than themselves. As they do not have a complete jaw, they use their knotted body as a makeshift one, while simultaneously generating torque to pull off chunks of meat. Hagfish prefer to twist themselves in simple overhand and figure-of-eight knots, but they will sometimes tie the more complicated three-twist knot as well.

Based on fossil records, hagfish have not changed much for the past hundreds of millions of years. It seems their slime-oozing, knot-tying survival strategy has little need of improvement.

Harp sponge

Chondrocladia lyra

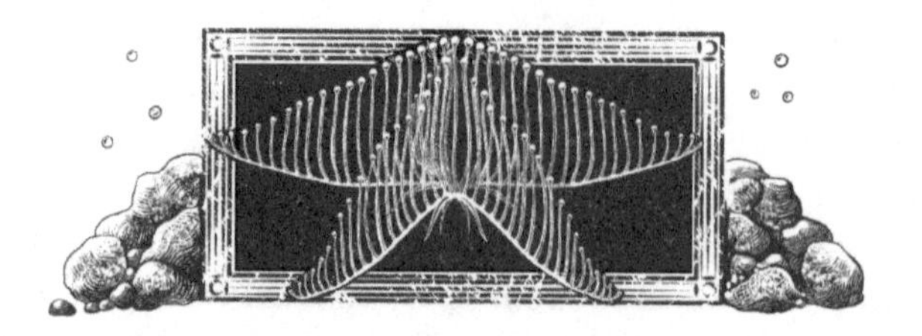

Is the harp sponge even an animal, or just one of those strange devices used for demonstrating physics phenomena in science museums? While it looks like it could be either, this bizarre, geometric construction is in fact as much a living critter as a dog or a mosquito. Harp sponges were only described in 2012, thanks to an expedition organized by the Monterey Bay Aquarium Research Institute; these sea creatures are not easy to come by without the help of a remotely operated vehicle, as they live at truly mind-boggling depths. The MBARI expedition found specimens at about 3,300m, but subsequently these organisms have also been discovered at depths of over 4,800m along the walls of the Mariana Trench.

Sponges—phylum Porifera—are the most primitive multicellular animals, dating back over 500 million years. They don't have a digestive, nervous, or circulatory system—and they are still doing just fine, thank you very much. They lack a skeleton, but prop themselves up using spicules—tiny structural elements that provide support and protection against predators. The spicules, shaped like snowflakes, commas, hearts, or arrows, look rather beautiful under the microscope, and are used for differentiating sponge species. For a long time, sponges have been considered very simple filter-feeders, with an appetite for bacteria and plant matter. The harp sponge, however, is a carnivore, and poses a threat to mobile animals—an impressive feat for a sessile, or unmoving, creature.

The ecology of this species is closely linked to its unusual structure. The harp sponge, *Chondrocladia lyra*, has been named for its branching, harp- or lyre-like appearance. The animal consists of a basic framework, called a vane, which supports a number of upright branches. The branches are covered in filaments: barbs and hooks that act like Velcro when catching prey. The harp sponge's victims, usually larvae and little crustaceans, are caught in the barbs, then transferred to the branches—either through their own struggle, or because of the filament retraction—where they are ensnared in membranous cavities and digested whole.

Hunting in this way is practical—the harp sponge is unable to pursue its prey more actively, since it is anchored to the sea floor with a root-like structure called the rhizoid. The spread-out vanes (normally one to six in number, and radially symmetrical, or star-shaped) maximize the chances of catching a drifting meal, because they increase the surface area exposed to currents.

This curious wide shape also helps with reproduction. In a feature that adds to their resemblance to a modern art sculpture, harp sponges proudly display a number of swollen balls, full of sperm, located at the end of each upright branch. The sperm they produce is packed in little parcels called spermatophores, which are then released, in the hope that they will encounter other harp sponges, or, more precisely, their lady parts. These, also ball-shaped, are placed halfway up the vertical branches; they are the sites of egg production, and, when a spermatophore serendipitously encounters one, fertilisation and egg maturation ensue.

Although, to untrained eyes, the harp sponge may resemble nothing more than a fancy toast rack, its minimalist structure demonstrates what it takes to survive in the extreme conditions of abyssal depths.

Herrings

Clupea spp.

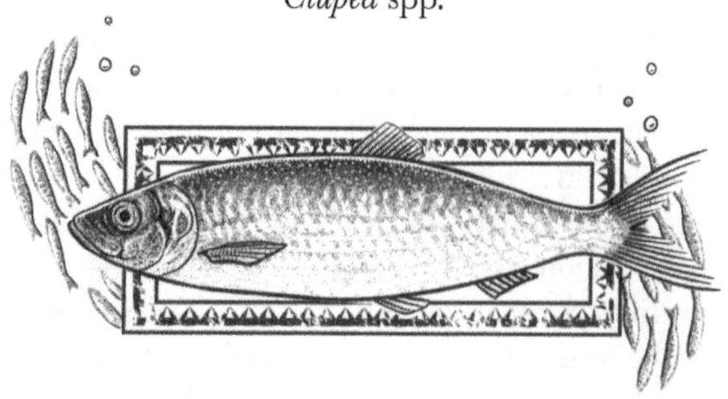

If you live in Europe, you've probably had at least a passing encounter with a herring; most likely in a culinary form. British kippers, German rollmops, Polish herring salad, or, for the brave, the Swedish fermented and pungent *surströmming*—for northern Europeans, herrings are as versatile as they are ubiquitous. There are three true herring species—Atlantic *Clupea harengus*, Pacific *C. pallasii*, and Araucanian *C. bentincki*, although a number of other fishes (both related and unrelated) are commonly called herrings. They are supremely gregarious creatures, living in schools of spectacular sizes: thousands, hundreds of thousands, and even millions of individuals. Their abundance led to their popularity as a food source: these oily, silvery fish have been eaten since 3000 BC; several towns and cities owe their existence to the trade in herrings—Amsterdam, Copenhagen, and Great Yarmouth to name a few.

Because of their commercial importance (in Europe they are known as the "silver of the sea"), herrings have been extensively researched in the context of optimized fishery outputs. Yet relatively little is known about some aspects of their biology, for instance social behavior.

One such comparatively under-studied element is fish noises. We don't normally think of fish as particularly loud creatures—if anything, quite the opposite. However, even though it might appear counterintuitive, herrings have a good sense of hearing, especially compared to other fishes. Their inner ear is connected to the air-filled swim bladder, which enhances the perception of sound and forms a hydro-acoustic detection system. Such complexity implies that hearing is important to the herring. But what are they listening out for?

One option could be the ultrasonic echolocation of dolphins and whales, who commonly feed on herrings. But while Atlantic herring may be able to snoop on predators, the Pacific herring's hearing doesn't quite span that high a frequency.

Another possibility is listening out for the noises made by other herrings. And it turns out that, indeed, such noises are made. How? Through their . . . derrieres. Herrings emit air bubbles via their anal openings (which are also connected to the swim bladder), to produce series of high-frequency pulses, ranging between 1.7 and 22kHz. Each series consists of tens or hundreds of pulses, and can last from a few milliseconds to a few seconds. These anally emitted sounds have been dubbed Fast Repetitive Ticks, or, for short, FRTs. The FRTs are likely used in interactions between conspecifics, a finding that won two international teams the Ig Nobel Prize in Biology. In 2004, Ben Wilson, Lawrence Dill, and Robert Batty, as well as Magnus Whalberg and Hakan Westerberg, shared the trophy for "showing that herrings apparently communicate by farting." They reached this conclusion based on three key findings. Firstly, the high frequency of the FRTs is audible to herrings, but not to most other species. Secondly, the greater the number of herrings placed in a tank, the greater the number of pulses emitted—disproportionally so, indicating that the animals engage in communal flatulence. Finally, herrings tend to FRT when it is dark, when they cannot see each other—the noises are very much absent during the day, when visual cues are likely to be their main mode of communication. While direct evidence is still absent, the authors suspect that thanks to this acoustic messaging, schools can stick together after dark.

Thunderous bum-pulses coming from multiple fish are a risky business, though, as they might potentially be overheard by the wrong crowd. Indeed, the specific characteristics of herring gases, such as frequency or duration, could be used by fisheries to pursue shoals more efficiently. Loud farting evidently comes at a cost.

Immortal jellyfish

Turritopsis dohrnii

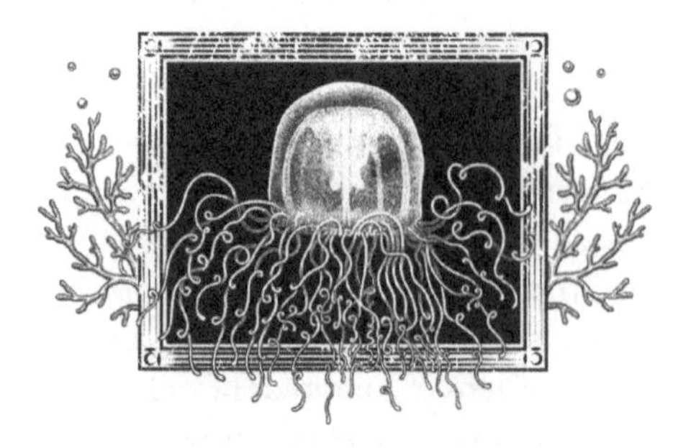

Who wants to live forever? Rock stars and jellyfish, that's who! But while the biology of the former gets in the way, the latter have it all sorted out. At least the immortal jellyfish, *Turritopsis dohrnii*, do.

To understand the unique, death-defying powers of the jellyfish, we first need to look more broadly at the lives of cnidarians, a large and diverse group of rather primitive aquatic animals. Jellyfish, technically known as medusae, are merely a phase in the cnidarian life cycle.

Cnidaria start off, not very remarkably, as swimming larvae. These, upon planting themselves at an appropriate site, grow into polyps, which look a bit like vases ornately decorated with a rim of tentacles. The polyps are a sessile form: they stay permanently attached to a substrate, and focus on growing. When mature enough, they develop a stack of plate-like medusae, which pinch off and float into the big wide world. Once this process, called strobilation, is complete, the polyp bites the dust (well, sea floor), or regenerates to strobilate once more.

The newly formed medusae are the free-swimming, sexual life forms often seen stranded on beaches. In the middle of the umbrella-like bell is a mouth, surrounded by tentacles; the mouth also doubles as an anus, because jellyfish are very economical creatures when it comes to anatomy. During spawning, the male and female medusae will release sperm and eggs into the water; with a bit of luck these stumble upon each other, fertilize, and form larvae—and the cycle begins anew. In some species, the sperm float into the female's mouth to fertilize the eggs there, which lends itself to speculations on whether the cnidarian equivalent of *Deep Throat* is a romantic comedy or a family drama. Other jellyfish don't try so hard and simply stick to asexual reproduction by fission or regeneration from a body fragment. After reproduction, the medusa eventually dies. While this life cycle is interesting, it is not that unusual. Larva–polyp–medusa, these are the days of cnidarian lives (though some species skip a stage here or there); an organism hatches, grows, has adventures on the way, and then expires or is eaten.

Yet the immortal jellyfish came to the conclusion that dying of old age is overrated. This tiny medusa, with a bell the width of a split pea and a bright red stomach inside it, is able to revert into a polyp when under pressure—skipping the fertilization and larval stages. While this "ontogeny reversal"—de facto "ageing in reverse"—has been observed in a few other cnidarians, it has never occurred in an organism that has already reached sexual maturity. In the curious case of *Turritopsis dohrnii*, the jellyfish use a process called dedifferentiation to rejuvenate. How does it work?

Cells generally start off as jacks-of-all-trades, with the potential to specialize into a specific type as they mature—a process known as differentiation. Once differentiation occurs, they stick to their roles, and produce more cells with the same specialty. However, dedifferentiation reverses that process: fully specialized cells dedifferentiate into an unspecialized state, and then re-differentiate into a completely different specialization, thus turning a mobile, sexual medusa into a sessile, asexual polyp.

The immortal jellyfish turn on their reversed ageing mechanism in response to stress, either due to mechanical damage, changes in temperature or salinity, or lack of food. They transform—first the tentacles are reduced, then the body shrinks and the ability to swim is lost. Finally, the regressed animal becomes encysted, settles in the substrate and eventually starts growing as a polyp. In theory, this process could continue indefinitely (unless of course the animal is eaten), ensuring immortality. It's—almost—a kind of magic.

Marine iguana

Amblyrhynchus cristatus

"It is a hideous-looking creature, of a dirty black color, stupid, and sluggish in its movements," wrote Charles Darwin of the marine iguanas he saw in the Galápagos Islands. His opinion seems somewhat unfair, given this animal has taken on such a unique lifestyle: it is the only algae-eating, sea-going lizard in the world. As reptiles, marine iguanas are reliant on their environment for heat, and thus, from an energetic viewpoint, feeding in cold water is no mean feat. With these challenges in mind, let's break down Darwin's judgmental claim.

"Hideous-looking"? Well, beauty's in the eye of the beholder. Iguanas' flat snouts (the scientific name, *Amblyrhynchus cristatus*, comes from the Greek *amblus*, "blunt," and *rhynchus*, meaning "beak," followed by the Latin *cristatus* for "crested") are the perfect shape for scraping algae from rocks. Powerful, flattened tails propel the animals underwater in an undulating, eel-like movement. Long claws help them climb on top of rocks. All in all, a rather practical package, even if not the most attractive.

"Dirty black color"? First of all, black skin helps iguanas absorb sunlight more efficiently when they bask; secondly, some of the eleven subspecies are not that dull at all. Depending on their diets, iguanas—particularly males trying to entice partners—can be red, pink, turquoise, emerald, gray, ochre, or almost white. The bright red-and-green lizards from Española Island have even been nicknamed the "Christmas iguanas."

"Stupid"? Okay, Darwin might have a bit of a point here. While predators such as domestic dogs were introduced to the Galápagos almost 200 years ago, the lizards still have not developed any effective anti-predatory responses. Hardly surprising, as their own disputes are settled by bobbing heads while open-mouthed—admittedly, this is also not the most intellectual of looks.

"Sluggish"? Nobody who has seen David Attenborough's *Planet Earth II* showing young iguanas dodging Galápagos racer snakes and jumping on to rocks with the agility of a martial artist could possibly agree. But here's something really interesting—for a species with "marine" in its name, they seem strangely reluctant to go into water. Only the biggest 5 percent of marine iguanas dive to feed—the rest prefer not to get their crests wet. Instead, they graze on exposed algae, running into shallow water for a quick bite when the tide is low. Even going for a brief plunge can be a big energetic burden for an ectotherm, and iguanas need to bask in the sun to "charge their batteries." The Galápagos Islands may be tropical, but the cold currents surrounding the archipelago mean that iguanas will cool down by up to 10°C when going for a swim.

As the reptile saying goes: when the temperature's lower, the lizards are slower. The cold affects their metabolism and behavior; chewing and digestion are more protracted, while movement is sluggish (leaving them vulnerable to predation). Herein lies the marine iguana dilemma: do I spend more time foraging, or re-warming after the dip?

The youngest iguanas bypass this quandary in a less-than-appetizing way: for the first few months after hatching, they feed on the droppings of grown-ups, which provide the necessary bacterial flora to digest algae. Meanwhile, smaller adults, who lose heat faster than large ones, are restricted to foraging in zones where (and when) the algae is most exposed. It might therefore seem like being bigger is more advantageous—but there is a flip side.

During El Niño events, when warmer waters cause the dieback of the preferred algal species, marine iguanas face starvation. Mortality can be as high as 90 percent, with the largest lizards being at highest risk, as they cannot meet their calorific requirements during food shortages. Amazingly enough, to improve their chances of survival, marine iguanas developed the ability to shrink—they can decrease their body length by as much as a fifth over two years.

Perhaps, had Darwin visited now, he'd describe marine iguanas as "creatures doing remarkably well, considering"?

Mary River turtle

Elusor macrurus

If you lived in Australia in the 1960s and wanted to buy a pet turtle, you were likely to stumble upon small, grayish hatchlings sold in pet shops throughout Adelaide, Brisbane, Melbourne, and Sydney. These little reptiles were so widespread that they gained the nickname "penny turtles" or "pet-shop turtles"—yet despite their popularity with children, they were not known to the scientific community. It was only in 1994 that John Cann and John Legler first described the species, after spending twenty years analyzing captive individuals, museum specimens, and, eventually, tracking the turtles down in their native habitat of the Mary River in southeastern Queensland. The newly discovered Mary River turtle gained the scientific name *Elusor macrurus*, from the Latin *eludo*, "to escape notice," expressing frustration at the elusiveness of the animal, the Greek *makros*, meaning "long," and *oura*, "tail," in reference to its characteristically shaped tail.

Due to algal growth on their heads and shells, these Aussie turtles may sport a rather punk-like appearance. On their chins, there are several barbels—protruding sensory organs that help them detect food in water. However, it is the other end of the Mary River turtle that is the most interesting. Like a number of other aquatic, air-breathing species, it faces a challenge—how to obtain enough oxygen and use it as efficiently as possible without having to resurface all the time for a gulp of air. Amphibians bypass this challenge by breathing directly through their skin—a solution not suitable for turtles, whose skin is thicker, and who, additionally, are covered with an impermeable shell. This reptile, however, found its own way out—via the way out. It is able to breathe underwater through its . . . bottom. It does so using the cloaca, a handy all-in-one opening containing the end points of the urinary, digestive and reproductive tracts. The cloacae of aquatic turtles also include "cloacal bursae"—pouches lined with finger-like protrusions that allow gas exchange directly from the water. The cloacal breathing can account for about a quarter of the Mary River turtle's total oxygen uptake, but in other species, such as the Fitzroy River turtle, *Rheodytes leukops*, this proportion can increase to 70 percent. While the elusive *Elusors* can spend two and a half days without resurfacing, thanks to such versatile behinds, as well as a slowed-down metabolism in lower temperatures, some aquatic turtles are even able to hibernate underwater.

Unfortunately, the reliance on bimodal (bum-modal?) respiration means that the Mary River turtle is not faring well in polluted or dammed rivers with a lower oxygen content. The species, which suffers greatly because of habitat loss and degradation, is now the second most endangered turtle species in Australia, and one of the twenty-five most endangered turtles in the world. Ironically, the pet trade that led to the species' discovery was also a factor leading to its decline, with egg harvests reaching around 2,000 per year in the 1960s and 70s to meet pet-shop demand. Hatchlings and eggs also suffer from predation by feral dogs and foxes. To make things worse, Mary River turtles are slow to mature, and females don't breed until they reach the age of fifteen to twenty years. Unfortunately, recent surveys found very few young individuals, which means that these unique reptiles struggle to make it to breeding age. Hopefully, with increased habitat conservation efforts, the "bum-breathers," as the Aussies call them, will have a better chance of recovery.

Mimic octopus

Thaumoctopus mimicus

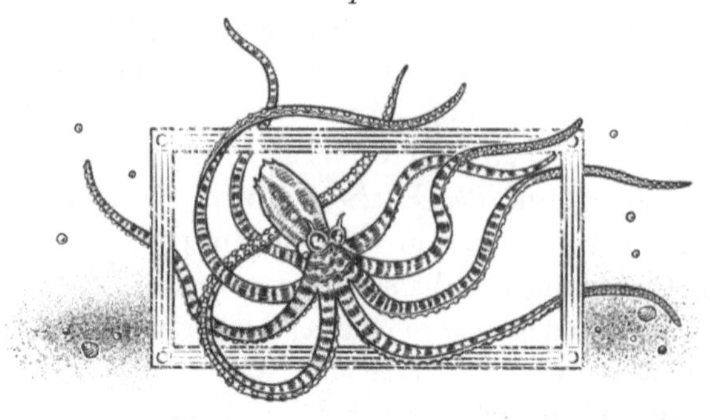

Octopuses are famously clever; they can use tools, crack puzzles, and they are also talented escape artists, constrained only by the size of their internal, hard beak. Yet while many cephalopods can change colors or patterns to find—or impersonate—mates (see Giant Australian cuttlefish, page 102, and Caribbean reef squid, page 166), and even emulate textures to match their environments, one talented mollusc one-ups them. The mimic octopus, *Thaumoctopus mimicus*, can imitate no fewer than thirteen different species—not just via skin color or pattern, but also the positioning and movement of the body. A sea snake? No problem—hide the body and six arms in the sand, wave the remaining two in a snake-like motion. A spiny lionfish? Sure thing—puff out the arms and float gracefully. A flounder? Here goes—arms together, body flattened, swim close to the sand. Don't feel like moving? There are sessile (immobile) options, such as a polychaete worm peeping from the sand, a sea squirt or a sponge. Only discovered in the late 1990s in the murky waters around Indonesia—and since reported as far as the Great Barrier Reef and the Red Sea—the smallish, stripy, white-and-brown octopus has astounded scientists with its rich repertoire of parodies.

Where did this knack for impressions come from? While most octopus species prefer to live in sheltered rocky habitats, mimic octopuses reside near river mouths, in silt and sand substrates. Their residence doesn't offer much cover, apart from the occasional hole in the ground, identified by poking and prodding (many octopuses are found with missing arm tips, probably from the time when the original owner of said hole was not particularly welcoming). Living in such an exposed habitat with lots of predators forced the mimic octopuses to resort to dynamic mimicry—or shapeshifting. But, in contrast to mimics that focus on just one look (see Orchid mantis, page 200), this mollusc has a whole range—and it dons the most appropriate disguise based on circumstances.

When pestered by small, territorial damselfish, the octopus becomes the venomous banded sea snake. The flatfish attire—likely mimicking poisonous soles abundant in the area—is handy for traveling at speed near the bottom of the sea. For swimming in the water column, the spiky, toxic lionfish impersonation provides most safety. While other octopuses prefer to be invisible, this species deliberately makes itself very conspicuous, indicating to predators that it may be noxious—and switching between looks in seconds.

Such a sophisticated evolutionary party trick requires a sophisticated nervous system. Unlike ours, the octopus' nervous system is not housed predominantly in the brain—two-thirds of its neurons are in nerve cords within the arms, which can perform reflex actions without input from the brain; they literally have a mind of their own.

Because of how complex their central nervous systems are, cephalopods have been granted the status of sentient beings, and in the UK are offered the same welfare protection as vertebrates. On the flip side, because of the mimic octopus' rarity and shyness, its conservation status is unknown. Its habitats are under threat from coastal run-off and mining and, as it gained media attention since being discovered, it is hugely desired by collectors. Few captive individuals survive: fishers use cyanide, copper sulphide, and other noxious chemicals to flush them out of their lairs, and aquarists rarely know how to look after them. By any sensible measure, surely the mimic octopus is worth far more alive in its natural habitat than dead after a brief stint in a poorly suited aquarium?

Olm

Proteus anguinus

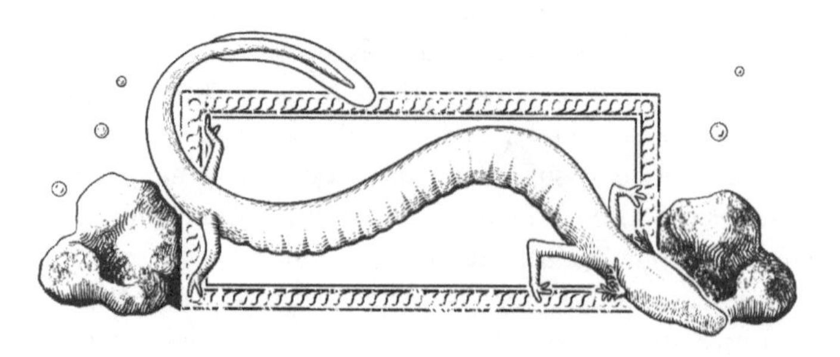

The olm, *Proteus anguinus*, is a 30cm-long, slender, flesh-colored amphibian found in the subterranean cave waters of the Western Balkans. Because of its unusual appearance (think infant-cum-lizard, or Gollum-cum-Smaug) it is locally known as a baby dragon or human fish—but, in reality, a more appropriate name would be the "why-bother newt." This remarkable animal is the embodiment of Ockham's razor, taking parsimony to soaring new heights.

Olms are troglobionts (from the Greek *troglos*, "cave," and *bios*, "life")—they live exclusively inside caves. Their silent, pitch-black, nutrient-poor environments necessitate equally stripped-down life-styles, body plans, and behaviors.

Growing up? Why bother? Olms are the Peter Pans of the vertebrate world. Being neotenous, they do not metamorphose like most amphibians, but instead the adults retain juvenile characteristics such as external gills, large heads and the ability to regenerate tails and limbs. Baby dragons reach sexual maturity in their teens; subsequently they do not reproduce often: a female will lay a few dozen eggs every twelve and a half years.

Sturdy legs, eyesight, skin coloration? Why bother? Olm limbs are thin and feeble, with fewer digits than other amphibians: only three toes on the front legs, and two on the hind. Olm larvae start off with normal eyes, but these atrophy and are overgrown by skin within a few months. Still, adults retain some photosensitivity (through these regressed eyes, as well as the skin) and will swim away from light if they encounter it. To make up for their lack of sight, these cave salamanders have an acute sense of smell, and can also pick up sound waves in water, vibrations from the ground, as well as electric and magnetic fields. The olms' whitish or pinkish appearance is down to their translucent, pigment-less skin—the only splash of color is provided by the red of the oxygen-rich blood flowing through their external gills. Still, baby dragons are able to produce the pigment melanin and will become darker if exposed to light. In fact, a black subspecies of olms, with shorter and stubbier bodies and with slightly better developed eyes, has also been described.

Moving around? Why bother? If you ever felt too lazy to get yourself something to eat, the olm might be your spirit animal. Olms are supremely sedentary, and while most have territories totalling only a few square meters, one individual has been reported not to have budged from the exact same spot for over seven years. These sit-and-wait predators will eat crustaceans and other aquatic invertebrates that happen to find their way near their snouts. If nothing ventures close enough, though, baby dragons will happily fast for several years.

Dying? Why bother? Olms, the largest vertebrate troglobionts in the world, seem to have found a key to eternal youth: their life spans are estimated at over a century, and, what's more, they do not show signs of senescence as they age. Are the cool ground waters of the Dinaric Alps a fountain of youth? More likely everything runs at a slower pace in the cold, dark, subterranean habitat, and these cave salamanders, with their unfussy attitude and high tolerance to low oxygen levels, fit right in.

Studying olms is not easy, since a lot of the caverns they inhabit require dangerous cave diving, or are downright inaccessible. Moreover, the marking methodology used for other amphibians—nicking a tiny bit of tail or toe—does not work on olms, as the clipped body parts soon regenerate. Thankfully, researchers are able to survey baby dragons using the DNA they shed in the water, for example from skin or feces. This technique, called eDNA, or environmental DNA, can confirm the presence of olms just from cave water samples.

Turns out that scientists can also play the "why bother" game.

Peacock mantis shrimp

Odontodactylus scyllarus

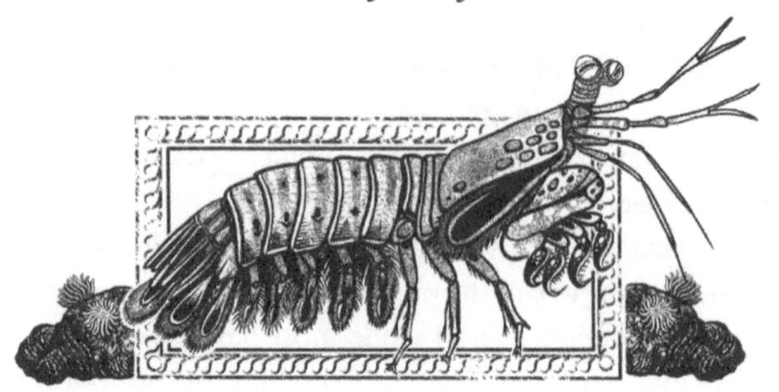

The Dactyl Club is not a new hip venue in Soho. Pretty much the opposite—it's a deadly underwater weapon wielded by a crustacean. The crustacean in question, the peacock mantis shrimp (*Odontodactylus scyllarus*), has a bright green body, orange legs, a spotted red head, and resides in tropical Indo-Pacific waters.

Like the elephant shrews or the mole-rats, mantis shrimps are among the many creatures named after other animals—while not actually being either. They belong to their own order, Stomatopoda (around 450 species in total)—not the same as shrimps, though both are in the class Malacostraca, and vaguely resemble one another in shape. The mantis part of the name comes from the way the stomatopods hunt: they are ambush predators, staying put and then rapidly launching their forelimbs at the passing prey—but here the similarities end.

The second pair of stomatopod thoracic limbs, menacingly named "raptorial appendages," is equipped with powerful weaponry adapted for close-range combat—and the forearms come in two varieties, depending on species. The spearers have a set of barbed spikes, with which they impale soft-bodied prey, such as shrimps or fish. The smashers, feeding on hard-bodied animals (snails, crabs), use a pair of bludgeons to shatter shells or exoskeletons. These hammer-like devices are the aforementioned dactyl clubs—"dactyl" means "finger" or, in this case, the terminal limb segment. Smashers tend to be diurnal, prefer clear waters and live in coral—they use their appendages to excavate a dwelling—while spearers inhabit murky environments and are more nocturnal.

The peacock mantis shrimp can smash using the heel of the dactyl, or spear using its tip—and, even though its body measures only up to 18cm, it can certainly deliver a heavy blow. At accelerations of 10,400*g*, their punches are as forceful as a .22 bullet; the spring-loaded strikes launch at speeds of 23m/s. The stupendous accelerations are achieved via power amplification mechanisms involving elastic springs, latches, and levers. Mantis shrimp thumps are so powerful that they create tiny cavitation bubbles between the club and the prey, which then collapse so violently that they produce a noise, heat, and a flash of light, as well as delivering an aftershock half as strong as the initial impact force. The poor snail or crab is first battered with the dactyl clubs, and then with the bubbles they created.

Such formidable weapons can withstand delivering thousands of energetic blows. This is thanks to their composite structure: they consist of three layers of varying flexibility, with the outermost one made of a very tough mineral called hydroxyapatite. And if all goes pear-shaped and the mantis shrimp loses an arm, it will grow a new one during the next moult.

With great power comes great responsibility, so to gauge the direction and accuracy of such a mighty punch, mantis shrimp eyes are among the most complex in the animal kingdom. While the spearers' vision is adjusted for hunting in dim light and at short distances, smashers' eyesight is incredibly acute. Their two compound eyes are divided into three parts, allowing them to focus light with three separate regions. The result? Independent, trinocular vision in each stalked, turning eye. While humans have three photoreceptors (red, green, and blue), the peacock mantis shrimp trumps that with twelve. They are able to detect circular polarized light—something we need 3D glasses for—and see in the UV range. The skill is presumably handy for communicating with other stomatopods via the UV patterns reflecting off their bodies; the males and females are polarized differently, for instance on antennal scales.

Even without their remarkable eyes, mantis shrimps can still detect each other via smell. They can be very territorial, so quick detection of a rival helps to avoid confrontation. Still, when it comes to fisticuffs over domestic disputes with members of their own species, they are mindful of establishing ground rules—and do pull their punches.

Pearlfish

family Carapidae

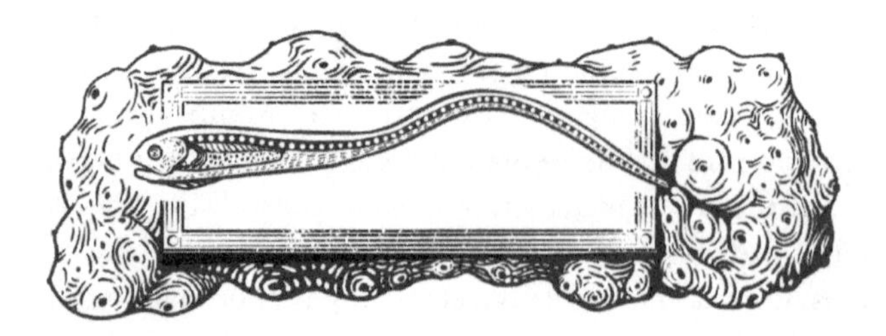

Location, location, location—it's just as key in the sea as it is on land. For pearlfish, small, thin, scale-less, eel-like fish from the family Carapidae, the prime residential spot is perhaps a bit different from ours: they live inside sea cucumbers" (see page 138) butts.

What makes a marine invertebrate's backside so liveable? Among other bum functions, sea cucumbers use theirs for breathing (similarly to Mary River turtles, page 118). They pass water through so-called "respiratory trees," organs that sit just inside the anus and extract oxygen. Because the behinds are so well ventilated, cozy, and soft, they can't help but attract visitors. Pearlfish find appropriate sea cucumber species via chemical cues, and they probably tell the back from the front by searching for the current produced by respiration. They enter their new lodgings either head-first, or, more often, rear-first, with the help of their long, tapering tails. If the sea cucumber refuses to cooperate—after all, not everyone wants to have a fish up their bum—and shuts the anal opening, the pearlfish wait: at some point the invertebrate has to breathe, and the fish seize the opportunity to break and enter.

While sea cucumber entrails are toxic to most animals, pearlfish appear to tolerate them rather well. Once inside, the fish is a tidy tenant—since its own backside is positioned right under its chin, all it needs to do to visit the bathroom is stick its head out of its host's anus.

Pearlfish may occupy a sea cucumber studio, or do a house share—it is not uncommon to find male–female pairings in one accommodation. Indeed, some fish couples decide to start a family inside their new-found love nest (or cheap cucumber motel?), and will mate and spawn in the safety of a derriere. The larvae are then usually released when the host breathes out, and they become part of plankton.

Conveniently, pearlfish are able to emit sounds by muscular drumming on their swim bladders—and do so when a cucumber they enter already has a lodger. The knock-like sounds help determine the sex and the species of the visitor, and can be heard outside their invertebrate homes. This ability to knock at the back door comes in handy, as certain pearlfish are more territorial and will attack intruders invading their private butt-space. Poor sea cucumbers may therefore not only have new life being created inside them, but they are also the sites of fights to the death, with the winner feasting on the remains of the loser. Still, some pearlfish are perfectly happy to have housemates, even from a different species, and the record-holding sea cucumber contained a whole dormitory—a whopping fifteen fish!

Pearlfish will also live in other invertebrates such as starfish, sea squirts or clams. Several species are altogether free-living. Those of the genera *Carapus* and *Onuxodon* are commensal, which means that they make use of residential hosts, but do not harm them. They exploit them as an abode, but emerge to go hunting, usually for other fish and small crustaceans—making use of their sizeable teeth and jaws. However, others, from the genus *Encheliophis*, treat their sea cucumbers like bed and breakfast, and feast on their insides. Their smaller teeth are perfect for nibbling on the soft entrails of their hosts, particularly their gonads. Unsurprisingly, some sea cucumber species have developed special defenses to protect themselves from these parasitic pearlfish, and evolved a biological equivalent of the medieval chastity belt: anal teeth!

Piure sea squirt

Pyura chilensis

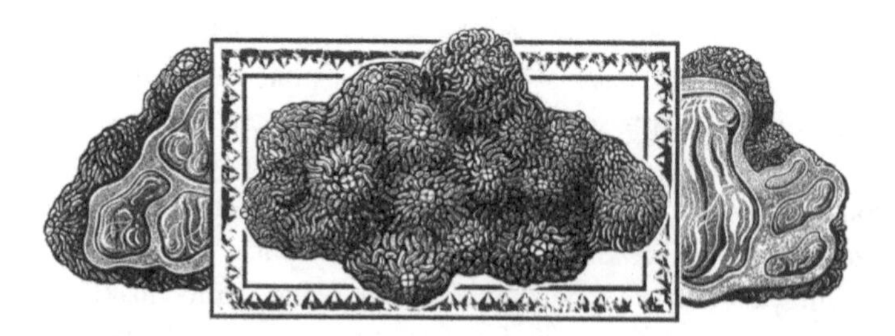

Imagine the horror of cutting open a rock, only to find bright red, bleeding organs inside. Thankfully, it doesn't mean you've been teleported into the world of some cheap gory slasher—you have merely cut open a sea squirt, *Pyura chilensis*.

Most sea squirts (subphylum Tunicata) are rather attractive; they dazzle with their wonderful selection of colors and shapes—some glow in the dark, some are embellished with patterns or bumps. In short, they are what makes all the under-the-sea Disney movies magical. But *P. chilensis*, also known by its Spanish name *piure*, clearly did not get the tunicate dress code memo. Rather than taking after its pals the "sea violet," "sea bluebell," or "sea peach" (who, frankly, sound like the underwater incarnations of My Little Pony characters), it looks more like a meat rock.

Whatever their outfits, sea squirts are built in a similar way. Because they are filter feeders, they have two tubes, or siphons—one to let the water and food in, and one to let water and waste out. The oral siphon is always placed upstream of the toilet one, to avoid a crap meal. Tunicates' bodies are (as you might guess from the name) encased in a tunic—a covering that may be rigid and cartilage-like, delicate and soft, or gelatinous and see-through. One of its main components is cellulose, which is unusual for two reasons. Firstly, cellulose is mostly found in plants; tunicates are the only animals able to synthesize it. Secondly, and uniquely for invertebrates, the tunic grows with the animal, eliminating the need to periodically shed the outer casing.

The tunic of the *piure* is thick, wrinkled and covered in sand and mud, giving it its rock-like appearance. Underneath, the body of an individual *P. chilensis* is blood-red and about 5cm long, with siphons discreetly poking out of the rocky surface. These living boulders inhabit rocky sea bottoms, either solo or in massive clumps, from a few to a few thousand individuals—giving the impression of tomatoes set in concrete. Found in the coastal areas of Chile and Peru, the species is often gathered by humans; it's a popular snack, eaten both cooked and raw, and is also considered (unfoundedly) an aphrodisiac. Currently, however, there is growing concern about the toxicity of *piure*: its efficient filtering can result in accumulating pollutants, and its body contains surprisingly high concentrations of metals, such as iron, titanium, or vanadium, probably used as a deterrent against predators. Despite this risk, this sea squirt is overharvested across its range, and consumed domestically, as well as exported to Sweden and Japan. The excessive collection of adults has posed a threat to the species, as recovery of harvested areas is very slow.

To make little sea squirts, *P. chilensis* have two options—finding a partner or going solo. They begin their adult lives as males and, after reaching the size of over 1cm, become hermaphroditic. These tunicates can either squirt out the gametes into the sea water, and then keep their siphons crossed that they find another gamete in the big blue ocean, or successfully self-fertilize. The former is a more common strategy, but both produce larvae. The larval stage is the only time in the life of a *piure* when it can move around—albeit only for a short period (12–24 hours), after which it settles, cementing itself to a substrate (or other sea squirts) before metamorphosing and setting its fate in stone. Because the motile youngsters are so short-lived, *P. chilensis* are not great dispersers.

The tadpole-like larva reveals a family secret. It does not resemble the barrel-like adult form; instead, it has a long, muscular tail, a hollow nerve tube and a notochord, just like hagfish (see page 108), frogs or humans. The notochord holds the clue: sea squirts are chordates, just like us! In fact, they are the invertebrates most closely related to people. When searching for ancestors, leave no stone unturned, especially not one pockmarked with red siphons.

Platypus

Ornithorhynchus anatinus

To say that the platypus, *Ornithorhynchus anatinus*, looks weird would be the understatement of the century. This animal is a glorious amalgamation of mammalian, avian, and reptilian features in one iconic, duck-billed, semi-aquatic, 50cm-long package.

Along with five species of echidna, platypuses are monotremes—egg-laying mammals. Monotreme means "one hole," and refers to their cloaca, the "one hole" usually found at the back end of birds and reptiles. Like placental mammals, platypuses have fur and produce milk. Like reptiles, they lay eggs, and the males are equipped with spurs containing venom akin to that of snakes. Like birds and mammals, they maintain a steady body temperature—though, at 31–32°C, it is much lower than that of their marsupial and placental relations.

Platypuses hunt for invertebrates in murky rivers and streams, with closed eyes, ears, and nostrils. To find food, they use the electro-sensory system in their bills, detecting the electric fields generated by the contracting muscles of their prey—worms, crustaceans, and insect larvae. Like birds, adult platypuses lack teeth, instead using horny plates to grind food. Even more unusually, they lack a stomach—all that is left of it is a widening at the end of the oesophagus, right before it connects with the intestine. Their thick fur is not only waterproof, but also biofluorescent, glowing blue and green under UV light. Platypus' eponymous feet are webbed (the name comes from the Greek words for "wide/flat feet"), and the extra-prominent webbing of its powerful front paws is folded during walking; to protect it, platypuses walk on their knuckles. Since their legs are on the sides of their body rather than underneath, the platypus gait is very reptilian.

How do such strange creatures come about? The process is not quite as simple as introducing Mr. Duck to Ms. Beaver, though, again, it involves tricks from many animal taxa. Platypus females have two ovaries, but, as in many birds and some reptiles, only the left one is functional (the right one is embryonic). Platypus males have testes inside their abdomens, while their two-pronged, spiny penis is stored in the cloaca, and only out when delivering sperm (peeing is done directly via the cloaca). Three weeks after a platypassionate rendezvous, the female lays a pair of leathery eggs the size of a Cadbury Mini Egg, and her 15mm-long young, delightfully called puggles, hatch ten days later.

To determine if a puggle is a boy or a girl, platypuses don't simply use two chromosomes. In most mammals, XX makes a female and XY a male, while in birds the system is flipped, with ZW creating a female, and ZZ—a male. In platypuses, a girl puggle is a whopping XXXXXXXXXX, while XYXYXYXYXY stands for a puggle boy. Instead of being content with just one pair of sex chromosomes, this species has five—and, what's more, one of those chromosomes carries a gene that is also found on the Z chromosome of birds.

Unlike placental mammals, which develop in the cushy and sterile environs of their mother's uterus, young puggles spend their early development in an underground burrow, in a nest of grotty wet leaves. To battle these unhygienic conditions, platypus milk contains a powerful antimicrobial component, the monotreme lactation protein. Because monotremes don't have nipples, the power drink is lapped up by the puggles directly from the mother's fur. Based on the insights from the platypus genome, it is now clear that milk production developed in mammalian ancestors before they stopped laying eggs; its purpose was to moisturise the parchment-like eggshells. Over time, early mammals started using milk as sustenance, and grew a placenta to nourish babies in lieu of egg yolks.

Thanks to the duck-billed oddity, we now know what came first —the mammal or the egg.

Racing stripe flatworm

Pseudoceros bifurcus

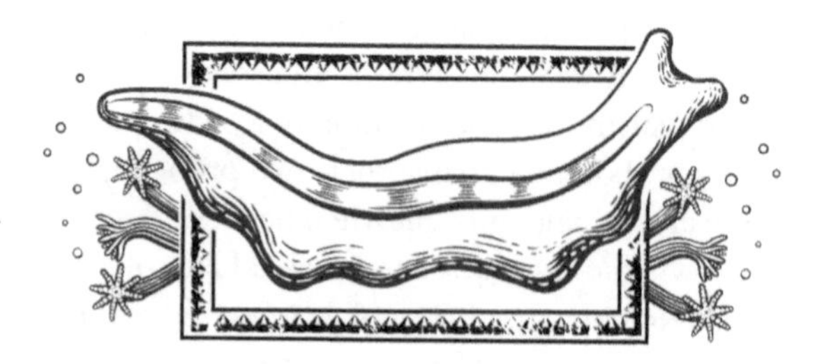

Racing stripe flatworms are gorgeous creatures: delicate, bright blue, with a striking black-white-and-orange stripe down their middle. Like all platyhelminths, they lack a respiratory system; their flat shape allows efficient gas exchange through diffusion. They resemble a 5cm piece of fine-looking haberdashery.

Obviously, such beauties should not have any problems finding a mate. Racing stripe flatworms are widespread in the tropical Indo-Pacific waters, and, to make dating matters easier, they are simultaneous hermaphrodites. What that means, apart from the fact that their preferred pronouns are "they"/"them," is that they have both male and female reproductive organs to make beautiful flatworm babies—and any conspecific has the potential to be The One.

But here is the catch: the responsibilities of being a mom and a dad are not equal in the marine flatworm world. The impregnated individual needs to expend energy on growing the new generation, whereas the one who sowed the wild oats swims away without a care in the world. On top of that, the mechanics of lovemaking are awkward, with the flatworms using their sharp penises to inject each other with sperm (akin to the traumatic insemination performed by bed bugs, see page 28)—after all, nothing says "sexy" like "hypodermic insemination." The sperm forms droplets under the partner's skin and migrates to the ovaries located on the animal's back. Consequently, one side of the couple ends up knocked up and wounded, while the other leaves and never calls back.

This begs the question of how the flatworms decide who gets which role. Well, flatworms are into sex games, and the most popular one is penis fencing. The rules are simple: whip out your penis, rear up, stab the other flatworm and don't get stabbed yourself. When you stab, inseminate. Each round is about 20–30 minutes long. There are no limits on the number of attempts, though the one who stabs first tends to be more successful at securing paternity. Uncompromising as it sounds, the winner moves on to fence with another flatworm, while the loser becomes a mom.

It might seem like for some flatworms love is a losing game, but, in fact, this lack of fairness does come with evolutionary perks. Trying to avoid a stabbing might be a test of stabber dexterity, driven by sexual selection: a more efficient stabber ensures better genes for the stabee's offspring. This explains why the copulation is a duel, rather than a more peaceful or passive affair.

Still, there are even worse scenarios than losing a penis battle, as proven by another flatworm, the 1.5mm-long, transparent *Macrostomum hystrix*. If the chances of finding a partner are slim, rather than letting the biological clock tick on, this lonely little flatworm opts for a forlorn attempt at self-fertilization: it stabs itself with its own needle-like penis. Usually in the head, because that's where, due to anatomical constraints, the penis reaches. The sperm then travels down to the tail region, to fertilize the eggs. Not an ideal tactic, but it beats not having any offspring at all. Plus, the term "selfie-stick" has never been a more desperate double-entendre.

Roving coral grouper

Plectropomus pessuliferus

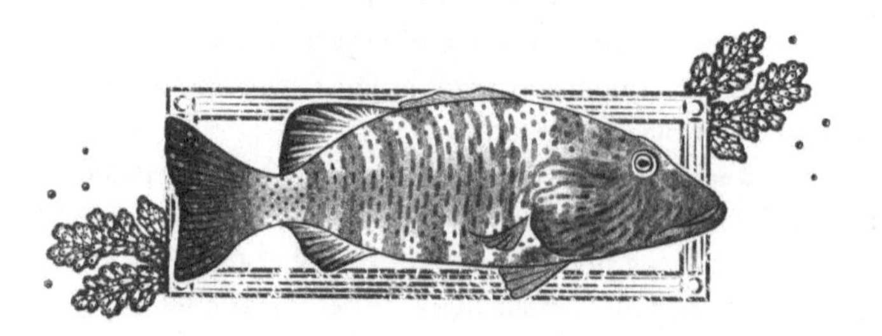

Every few years, the world is shaken by reports of "monster groupers"—massive, 400kg fish capable of swallowing a small shark in one gulp. While some species of groupers do reach such impressive sizes, the roving coral groupers, *Plectropomus pessuliferus*, over 1.2m in length, are just as much about the brains as they are about the brawn.

Groupers inhabit coral reefs in the Indo-Pacific region. They are carnivores and mainly feed on fish and crustaceans, hunting for their meals in open waters. To avoid being eaten, reef fish hide between the corals, where bulky groupers cannot access them. This might seem like the end of a hunt for the day—however, hungry groupers are not ones to give up easily. Instead of admitting defeat, or waiting around for the prey to come out, they devise cunning pursuit tactics to flush the fish out of their hidey-holes.

A grouper hunting plan involves three simple steps: 1) recruiting partners; 2) showing them what to do; and 3) splitting the bounty.

A good partner for a cooperative hunting spree has a specific set of skills, complementary to the grouper's. A moray eel, such as *Gymnothorax javanicus*, is a great candidate. Slender-bodied morays hunt in reef crevasses; they are able to squeeze through narrow spaces and flush out prey inaccessible to the groupers. Napoleon wrasses (*Cheilinus undulatus*) also make suitable expedition partners—though for a different reason. They are able to use their massive jaws to smash the reef, or to suck out hidden prey. Either way, fish hiding in the reef flee into the open waters, where the groupers await. Pursuit partners don't need to be piscine—octopuses, who, like morays, are experts at squeezing into tight spots, have also been known to join collaborative hunts.

Until recently, cooperative hunting where each partner plays a different role had only been seen in a handful of animals, all of them mammals or birds. This complex social behavior requires good communication and a clear signal for initiating the hunt. Groupers recruit morays by using specialized gestures—they swim up to the sheltering moray, arrange themselves in full view, and do a vigorous shimmy: a "Giddyup, pardner!" in fish-speak. The shimmy can be repeated multiple times if the moray is slow to join the hunt, or is distracted halfway through. Groupers also indicate what it is that they want their associates to focus on, using so-called "referential gestures"—an equivalent to humans pointing to something. They will do a "headstand," swimming vertically, head-down, above the place where escaped fish have been hiding after an unsuccessful pursuit. The headstands are always performed in the presence of potential hunting partners, and more often than not gain their attention and elicit a response. The ability to use referential signals is seen as a huge cognitive feat, again previously only attributed to birds and mammals, so noticing super groupers pointing to items of interest has led scientists to rethink the mechanisms of animal cognition.

Finally, for the hunt to be mutually successful, both parties must be better off at the end of the partnered prowl than they would have been going it alone. And indeed, it's been shown that cooperative hunters catch more prey—in the case of groupers, almost five times as many as during their lone stints. Because both groupers and morays tend to swallow their prey whole, there is little room for squabbles. Fish who want to grab more than their fair share should watch out when hunting alongside an octopus, though—these molluscs have been known to punch an uncooperative or greedy partner.

Sacoglossan sea slugs

Elysia marginata, Elysia atroviridis

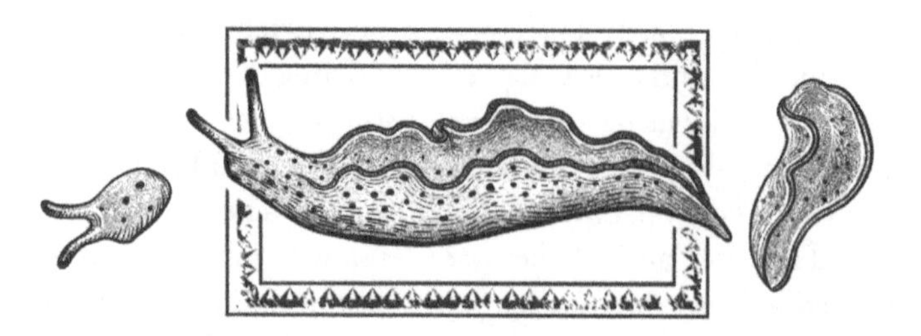

Many animals love basking in the sun—but could any animals actually be solar-powered? Unlikely as it sounds, yes.

Sacoglossan sea slugs are a group of sea-dwelling molluscs; two species, *Elysia marginata* and *Elysia atroviridis*, look like green, winged worms with a pair of tentacles on their heads. These sea slugs feed on algae—not an unusual diet for marine creatures. However, sacoglossans exploit algae in more ways than one: not only do they eat them, they are also able to steal their chloroplasts, the part of a plant cell that enables photosynthesis. The slugs use the biggest tooth of their radula (a sluggy food-scraper mouthpart) to pierce algal cells; they then digest most of the cell contents, but leave the chloroplasts, which are incorporated into the molluscs' own digestive gland cells. While this incorporation is a feat in itself, what's even more amazing is that the *Elysias* manage to keep the chloroplasts photosynthetically active and working for them for months on end. The chloroplasts produce carbon compounds, which the sea slugs then feed on. Chloroplast robbery is called kleptoplasty, and is usually seen in primitive unicellular organisms, the protists. Kleptoplasty is extremely rare in the animal kingdom—the only other creatures capable of it are some marine flatworms.

So, would having active chloroplasts inside their bodies make the sea slugs fat just from sitting in the sun? Not quite. Still, for *E. atroviridis*, the combination of food and strong light slows down weight loss throughout their lifetime—more so than the combo of food and weak light. These conditions also account for higher numbers of eggs, larger-sized larvae and a better survival rate of the offspring. Plastid snatching clearly pays off.

While *E. atroviridis* and *E. marginata* cannot sustain themselves entirely on a chloroplast-provisioned diet (unlike their cousin *E. chlorotica*, who can live off the chloroplasts for a year or so), they make up for it with an even more extraordinary talent. They can lose their heads—or, rather, their heads can lose their bodies.

This is not a metaphor—the two *Elysia* species are capable of autotomy, or self-amputation. While the phenomenon has been observed in a number of animals—lizards or salamanders losing their tails in the presence of a predator, for example—it has never been seen elsewhere in such a radical form. In the self-beheading process, the slug severs off around 80–85 percent of its body weight, including the heart and other organs, along a neat "neckline"—and the head wanders off on its own. The body is still alive for a few weeks, or even months, and the heartbeats, more and more faintly, up to the point of decomposition. The head, however, starts a new, solo life, and proceeds to grow a fresh body, in an act of extreme regeneration. The new bod is ready in under three weeks, complete with heart and all.

The process of self-amputation lasts several hours, which means that it is unlikely to be a predator defense—it simply takes too long. Instead, autotomy could be a way of getting rid of parasites; *E. atroviridis* affected by parasitic copepods (small crustaceans) can shed their bodies and regrow new, parasite-free ones. Some individuals have even been observed doing it twice!

It is hypothesised that the key to the head's survival for long enough to regrow the body is kleptoplasty. The stolen chloroplasts are able to provide photosynthetic sustenance until the slugs can once again digest food—a neat trick to enable them to leave it all behind.

Sea cucumbers

class *Holothuroidea*

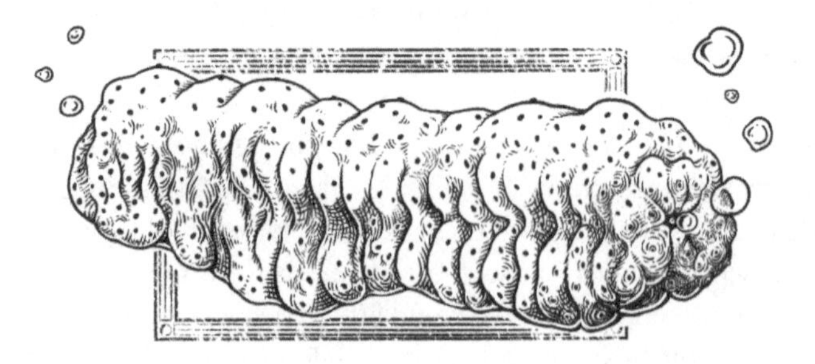

Sea cucumbers, like their plant namesakes, are generally longish in shape, lack a brain and don't move particularly fast. This, however, is where the similarities stop. Sea cucumbers, classified as echinoderms, are relatives of starfish and sea urchins. There are over 1,700 species, most of which are within the 10–30cm range, though the smallest measure only a few millimeters, while the largest reach over 3m in length. Most are tubular, some resemble worms or snakes, while others (the "sea apples") are almost round. Although they might not win sprint races against most animals, sea cucumbers can move—some using little tube feet, some through appendages modified to form fins or sails, others by crawling. And while they don't have brains, they still possess a simple nervous system, which makes them sensitive to touch and the presence of light.

This might not seem like much, but there really are very few creatures as cool as a sea cucumber. Firstly, thanks to its body being largely composed of so-called "catch connective tissue"—collagenous tissue that can readily change its mechanical properties—the sea cucumber can exist in three states of elasticity. The first is the standard state; the second, triggered when the animal is touched, is the stiff state. The third one becomes apparent when the sea cucumber is squeezed hard—it's the soft state, which allows the creature to pretty much liquefy its body (useful for pouring oneself through tight spaces). All these states are fully reversible, and the collagen can make cucumber bodies firm again.

The ability to change state comes in handy when confronting predators, as some species have a nifty defense mechanism: evisceration—

expelling their sticky and toxic viscera at the attacker. The combination of stiff and soft states helps control which bits stay, and which go—thankfully, the internal organs grow back within a few weeks. Similarly, if conditions are not favorable (i.e. the water or food is not right), sea cucumbers will eviscerate or reabsorb their gonads, perhaps in an act of protest that the cruel world is just not worth bringing babies into.

If the conditions are right, however, these long echinoderms will go ahead and breed. It is not a particularly intimate affair: male cucumbers release sperm, female cucumbers release eggs, and both sides hope for the best—that is, for the zygotes to form somewhere in the water. However, at least ten species can reproduce asexually, through transverse fission, or splitting in half. Both the front and the back start twisting, each in a different direction, until the middle is thinner and thinner, and finally splits. Most of the time both sides survive, although the front end is less likely to make it through.

As if all this wasn't enough, sea cucumbers are incredibly important to marine ecosystems. As sediment feeders, they ingest sand, digest the detritus, and return clean sand back to the environment—very much like sea earthworms. They also mix layers of sediment (improving its oxygen content), help recycle nutrients, reduce water acidity (which helps corals grow) and increase biodiversity. Phew—some responsibility for a creature that doesn't even have a brain.

Because of their regenerative powers, sea cucumbers are a sought-after culinary and medicinal item in China; due to their phallic shape and the ability to ooze internal organs, they are considered an aphrodisiac. Additionally, serving sea cucumbers at banquets is a means of displaying wealth, since high-value species can exceed $1,800 per kilo. While they can be found globally, the highest number of species inhabit the Asia-Pacific region; currently over 90 percent of tropical coastlines worldwide belong to countries exporting sea cucumbers to China. Unfortunately, they are very much overharvested, with 38 percent of sea cucumber fisheries overfished, and a further 20 percent depleted as of 2011. Meanwhile, demand is steadily growing, with around 200 million animals harvested each year, and both common and endangered species consumed. Consequently, these fascinating delicacies might all be eaten before we have a chance to properly study their superpowers.

Sea walnut

Mnemiopsis leidyi

It's small, it's squidgy, it's shiny—and it wreaks havoc across marine habitats. The humble sea walnut, *Mnemiopsis leidyi*, an iridescent, 5cm gelatinous blob, looks less like a deadly invader and more like a cheap novelty toy sold at a funfair. This unassuming animal is a ctenophore, or a comb jelly—not a jellyfish, if you please; they belong to completely different phyla.

Only around 200 ctenophore species have been described so far. Ctenophores are a bit more complex than sponges, and just about as complex as jellyfish—but simpler than all other animals. Like true jellyfish, comb jellies are usually transparent, consist mainly of water, and are mostly carnivorous, but they have an extra element: combs, or ctenes. These are eight rows of small iridescent hairs, called cilia, which propel the animal like tiny oars.

While swimming around in sea water using light-refracting combs is kind of cool, ctenophores' real claim to fame is being the oldest animal group with a nervous system—and perhaps the earliest developed creatures altogether. The scientific world is split on that matter, with some researchers backing ctenophores, and others the sponges (Porifera) for the trophy of the most ancient animals.

The seemingly unremarkable sea walnut has some pretty interesting features—but nothing that, on the surface, would foreshadow disaster. Like other comb jellies, it does not sting; it is bioluminescent, producing light by using a luciferase enzyme and the light-emitting substrate luciferin. While other comb jellies trail tentacles with long sticky cells to catch food, the sea walnut has two muscular, lip-like lobes on both sides of the oral cavity, and it hunts by swallowing prey whole, like a see-through, floating mouth. Once it is done eating, it makes use of a transient anus, a bumhole that appears for defecation (about once an hour), and disappears when not in use. It was initially believed that there were three different sea walnuts in the genus *Mnemiopsis*; currently scientists agree that they are all a single, though very diverse, species (once, twice, three times a *leidyi*).

Native to the western Atlantic Ocean, this comb jelly made its way to the Black Sea in the 1980s, brought with ballast water on ships—a well-known transmission route for aquatic invaders. Sea walnuts are flexible when it comes to living conditions, being tolerant to pollution, as well as a wide range of temperatures and salinities. They feed on small crustaceans, fish eggs and larvae, something that their new home had plenty of. Their resilience, voracious appetite and a paucity of predators led to a boom that turned the Black Sea into a sea-walnut ball pit, with levels soon reaching 300 ctenophores per cubed meter. What followed was a disruption of the already delicate ecosystem, a collapse of fisheries and a spread to neighboring seas. As hermaphrodites, sea walnuts can self-fertilize and produce 2,000 eggs daily—all it takes is one individual. Over the subsequent few decades, they have become the scourge of the seven seas: Black, Azov, Caspian, Marmara, Mediterranean, North, and Baltic, eating and outcompeting fish stocks important for humans and native wildlife. Following the arrival of the insatiable alien comb jelly, five species of sturgeon, as well as the already very fragile Caspian seal (*Pusa caspica*), declined further. Consequently, the sea walnut gained a dishonorable spot in the IUCN's list of 100 worst invaders in the world.

Almost twenty years after the first invasion, another non-native ctenophore, *Beroe ovata*, made it to the Black Sea. Thankfully, it delighted in eating sea walnuts—and offered a bit of a relief to Black Sea fishes. While in the Azov and North seas the low winter temperatures kill off comb jellies seasonally, reinfestations still happen on a regular basis. Fortunately, a convention signed in 2016 obliges all ships to destroy any beasts present in ballast water, which reduces the chances of future invasions. After all, it is worth remembering that even the smallest, squelchiest being can have a huge impact on ecosystems.

Spiny dye murex

Bolinus brandaris

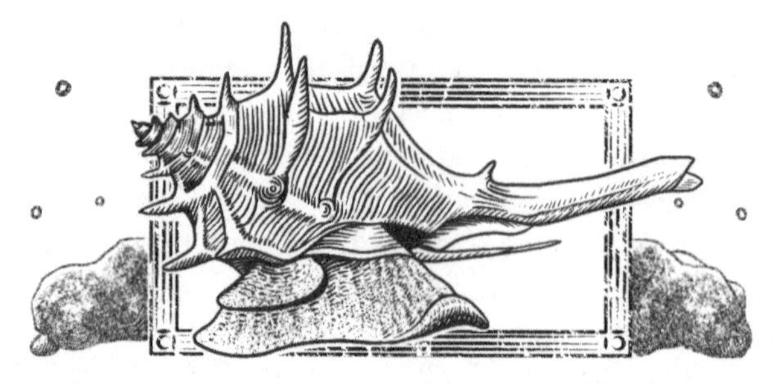

Can paint increase penis length? The short answer is: yes. The long answer is: yes, if you are a marine snail, and if you are exposed to paint containing anti-fouling agents. And—before anyone gets any ideas about having their penis painted—it is *not* a good thing.

Since the earliest days of maritime travel, naval engineers have dealt with the issue of biofouling: the accumulation of microorganisms, plants, or small animals on the surfaces of vessels. The build-up of various aquatic organisms on the hull, pipework, grates, and other equipment is bad news for the ship; it can cause damage, pose safety problems, or slow down the vessel by increasing its drag (which leads to larger fuel consumption and thus higher operating costs). Unsurprisingly, the shipping industry has been trying all kinds of approaches to prevent critters from choosing vessels as their mobile homes, from nailing sheets of copper to the hulls in the early days, to the most modern shark-skin mimicking coatings. One solution commonly used in the 1960s and 70s was anti-fouling paint containing organotin compounds with biocidal properties, such as tributyltin and triphenyltin. These, when leached into the environment, not only prevented the attachment of wildlife to boats, but also affected non-target species.

An animal acutely impacted by anti-fouling paint was the spiny dye murex, *Bolinus brandaris*, a sea snail inhabiting the Mediterranean and parts of the Atlantic. This spiky-shelled gastropod has been of commercial importance for millennia, having been used by the Phoenicians to produce the very highly prized "Tyrian purple" dye. The opulent pigment, extracted from the snail's hypobranchial (mucus-producing) gland, would not fade with time—and, because it required an expensive and complex production process, it could only be afforded by the wealthiest, becoming a symbol of status and power. In fact, the use of materials dyed with this rich color was restricted by sumptuary laws in Roman times, and by the fourth century AD only the emperor could "don the purple." Currently, the species is consumed around the Mediterranean, particularly in parts of Spain.

Although most terrestrial gastropods are hermaphroditic (see Banana slugs, page 18), neogastropoda, such as the spiny dye murex, are gonochoristic—that is, with separate male and female snails. While this already halves the chances of finding love, the sea snails have an even bigger problem. Low concentrations of organotin compounds leaching from anti-fouling paints have been causing the female snails to grow male genitalia—a phenomenon known as imposex, since the male bits are super-*imposed* on to the female ones. While some snail species can still reproduce under the multiple-organ circumstances, others, like the spiny dye murex, cannot, with the sterility of the females usually caused by the newly acquired vas deferens blocking the genital pore.

Imposex affects around 200 species, and is one of the best-documented and clearest examples of chemically induced endocrine disruption in animals. Even though the use of biocidal organotins has been restricted by the International Convention on the Control of Harmful Anti-fouling Systems on Ships, which came into force in late 2008, contamination can still be currently observed, particularly in shipyards, ports, and marinas. And, despite a decrease in the incidence of imposex-affected gastropods in some habitats as a consequence of legislative changes, a new problem arises: microplastics, which are ingested by the snails and find their way back into the stomachs of people consuming them.

The spiny dye murex, a highly sensitive species, is a very good bioindicator for environmental pollution. More broadly, the occurrence of imposex helps to assess how badly an area is affected by organotin compounds; the severity is measured using biomonitoring metrics such as the Vas Deferens Sequence Index, the Relative Penis Length Index and the Relative Penis Size Index. This must be one of the few circumstances in which some members of a species would prefer to possess a smaller penis.

Surinam toad

Pipa pipa

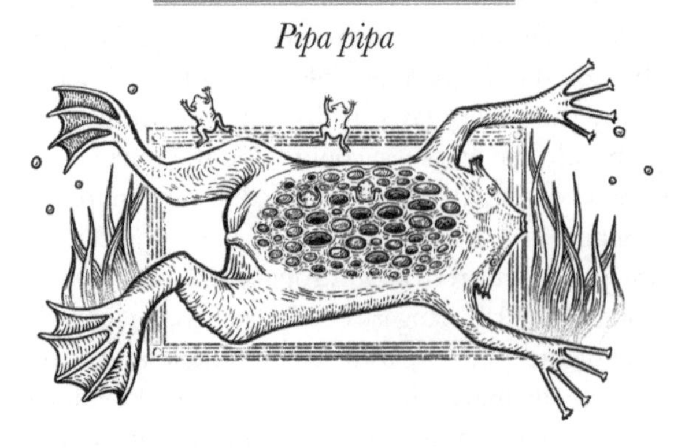

Are you suffering from trypophobia? That is, does the sight of multiple holes elicit discomfort in you? If so, you might want to skip this section.

You see, there is this toad. It's some 10–20cm long, weighs about half a kilo and resembles a rectangular pancake with limbs. The Surinam toad, *Pipa pipa,* is perhaps not the most handsome of the frog folk, looking more like a grayish or brown dried-up leaf than an amphibian, but thus far, nothing too alarming.

The species is found in tropical regions of South America. It is very much aquatic, residing at the bottom of murky, stagnant ponds and slow-flowing streams and rivers. While Surinam toads will scavenge given the opportunity, they particularly enjoy eating live prey, such as fish or invertebrates. However, they don't have teeth or tongues, and their eyes are rudimentary—so how do they hunt?

The answer lies in their sense of touch. These frogs (by the way, the distinction between toads and frogs is rather informal—toads are a type of frog) have two touch-tricks up their sleeve. The first is the lateral line, an organ usually found in fish, used for detecting motion, pressure, or vibration and translating it into electrical impulses. The second is their eerie fingers. While the toes on the hind legs are normal, frog-like, webbed toes, the forelimbs don't have webbing, but, instead, they have very long, witch-like fingers. Each of the fingers ends with its own fingers—which in turn have their own mini-fingers, fractal-style. These star-like appendages help the toad identify any movements of the water, giving away the location of potential prey.

How does hunting work? Surinam toads are ambush predators; they lie at the bottom of a pond, motionless, with their fingers outstretched, and wait. If an unsuspecting fish comes too close, the toad opens its mouth and—whoosh!—sucks it in. It uses its hands to shovel any parts of the prey still on the outside of its mouth. To create enough suction for this trap to work, the frog is able to compress its viscera (lungs, liver, stomach, etc.) and shove them toward the back of the body by a third of its length, to make space for an enlarged mouth cavity.

This suction-feeding mechanism is not even the strangest feature of the Surinam toad. What takes the prize is the puzzling fact that females give birth through their backs. During copulation, the male clutches the female in a love grasp called amplexus—he holds on to the female's back and fertilizes the eggs as she releases them. As this happens (and toad sex can take a while: twelve to twenty-four hours!), the female swims around, often doing somersaults—thus the fertilized eggs end up deposited on her back. Because of hormonal changes, the skin on the female's back swells, and the eggs sink into it, nesting in little pockets. The 40–120 eggs form an uneven honeycomb pattern, which is a common trigger for trypophobia.

The froglets transform inside their mom's back, undergoing all their developmental stages, from egg, through tadpole, to a fully formed youngling. Some two to three months after conception, when they are ready to start independent lives, they emerge as miniature Surinam toads. They do so by popping out of their mother's back, very much like limbed zits, and she then sheds the surplus skin. A true, dermatological miracle of life.

Tongue-eating louse

Cymothoa exigua

How is this for a conversation starter: "Have you ever thought what it would be like to have a trilobite for a tongue?" For several species of fish, this is not a theoretical question at all, thanks to small marine crustaceans called isopods.

Isopods resemble the now extinct trilobites: they are oval, have segmented exoskeletons and seven pairs of limbs. They are generally found in the water, although some isopods, such as woodlice (also known as slaters, pill bugs or roly-polys), prefer a terrestrial lifestyle. Many of the aquatic crustaceans are parasitic, and the family Cymothoidae in particular are known for their grisly habits. They are ectoparasites (which means they live on the surface of their hosts' bodies, rather than inside them), and usually target fish. These isopods will attach themselves to various sites on the body—skin, gills, fins, or mouth; sometimes they will even bore into the muscles. One species takes the gore a step further, though: it's the tongue-eating louse, *Cymothoa exigua*—and it is not difficult to guess what it specializes in.

The tongue-eating louse does pretty much what it says on the tin. When inside the mouth of its victim, this flesh-colored isopod will use its claws to sever the blood vessels to the tongue, causing its degeneration. Then, in a home-made attempt at organ transplantation, it attaches itself with its fourteen legs to the remaining stub and the floor of the mouth, becoming a nifty, living, legged replacement for the fish's tongue. Interestingly, the lice do a pretty good job as proxy tongues, since they are roughly the right size and fit cosily inside fish mouths. A fish tongue is not muscular like those of most other vertebrates—it is bony; its primary function is to secure food in place. During foraging, as the tongue presses against the roof of the mouth, it pushes the food against vomerine teeth (teeth growing out of the palate). Tongue-eating lice found in the mouths of snappers have been reported to show abrasions from these palate-teeth on their thoraces, indicating that they are doing a great job as tongues. What also speaks for them as living prosthetics is the fact that fish with louse tongues seem to be fairly healthy, if a bit thinner than unparasitised animals—after all, the isopods do feed on blood and fish mucus. Still, having a louse tongue is a better option than not having a tongue at all.

As if the tongue-eating business was not strange enough, the louse has another eccentric feature on its CV: protandrous hermaphroditism. Upon finding the right fish—and it is not too picky, having been spotted in at least eight different species in the warm waters of the Atlantic and eastern Pacific—the isopod moves in via the gills and sets up shop. But here is the twist: all tongue-eating lice start off as males, and only transform into females when they are over a centimeter long. If two lice find themselves in one fish, the smaller remains in the gills as a male, whereas the larger moves into the mouth and, apart from severing the tongue, it also undergoes a sex change. The females are twice the size of males, reaching over 3cm in length; they also have a marsupium—a handy pouch for keeping the eggs (several hundreds of them). The male generally remains in the gills, but pops into the fish mouth for conjugal visits. Yes, that's right: the fish unwittingly hosts a pair of copulating crustaceans *inside its own jaws*. Chew on that . . .

Water bears

phylum Tardigrada

On the whole, scientists are peaceful, friendly, unobtrusive beings. Yet over the past few centuries, a small group seems to be strangely fixated upon one morbid goal: to kill water bears—in the most extreme and elaborate ways possible.

What are water bears, and why would anyone want to kill them? Are they dangerous? Do they spread diseases? Have they done anything to deserve the wrath of researchers? Not at all. Water bears, also known as tardigrades or moss piglets, are minuscule invertebrates, averaging half a millimeter in length. There are some 1,300 species in the phylum Tardigrada, all looking like microscopic versions of eight-legged Michelin men. Though tiny, they have brains, intestines, gonads, and even little claws on their legs. These podgy, lumbering minibeasts are completely harmless to humans; they usually eat bacteria, but some are vegetarian and others carnivorous, preying on microinvertebrates even smaller than themselves. The only reason why scientists obsess about destroying them is because they like a challenge—as a phylum, tardigrades are the most indestructible animals on Earth (and beyond).

Water bears are found literally anywhere wet; in fresh water, marine habitats (from intertidal areas to depths of over 4.5km) or on damp bits of land—some moss, wet sand or leaf litter is ideal. However, this aquatic–terrestrial environment is prone to drying, and so tardigrades need a way to cope with its unpredictability.

When the going gets tough, moss piglets enter a state of suspended metabolism, called cryptobiosis—triggered by freezing, desiccation, absence of oxygen or high concentrations of chemicals.

To enter it properly, they form a dome shape called a tun—they shrink their bodies, suck in their legs and reduce surface area as much as possible. The tun slows down water loss and prevents organ damage during the subsequent drying process. Tardigrades then replace water with bioprotectants such as trehalose; they also beef up on other protective molecules, such as heat shock proteins (see Saharan silver ant, page 62) and damage suppressor proteins that shield DNA from radiation. Once liquid water becomes available again, water bears activate themselves, and pick up where they left off—like a "just add water" ready meal.

Cryptobiosis was first observed by Antonie van Leeuwenhoek, the inventor of the microscope, in 1702; tardigrades themselves were properly described by German zoologist Johann August Ephraim Goeze in 1773. Since then, scientists have been testing their limits. And, boy, do they have far-reaching limits. Even though their regular life span is a few months, when in tun they can survive without water for decades. They withstand temperatures ranging from a fraction above absolute zero (−272.8°C) to over 150°C. They tolerate astonishing pressures, and gamma rays and X-rays that would kill mammals hundreds of times over. While their amazing resilience is attributed to the metabolic standstill they undergo in the tun form, even hydrated, active moss piglets are more resilient than most animals; they can happily survive in temperature ranges of −196°C to 38°C, and pressures around a hundred atmospheres (equivalent to the pressure 1,000m underwater).

Because of their amazing hardiness, water bears became a model organism for space research. The latest chapter of "To Kill a Tardigrade" featured moss piglets sent to outer space to see how they handle vacuum (just fine, thank you very much) and the full blast of the sun's UV radiation (a handful still made it back OK). In 2019, panic ensued when it became apparent that an Israeli probe that had crashed into the moon was carrying tardigrades on board. Did humanity spill life—invincible life, at that—on the moon? To investigate if water bears could have survived the collision, some were loaded into a two-stage light gas gun and fired at sand targets in a vacuum chamber. Turns out that they can withstand impacts of up to 900m/s, or 1.14 gigapascals of shock pressure—but this is much less than the shock pressure of the crashing space probe. We have not created our own indomitable aliens after all. Yet.

Wattled jacana

Jacana jacana

"Don't be so picky! When will you settle down?"
Humans aren't the only ones faced with pressure to find a partner. However, the animal world is perhaps a bit simpler when it comes to who can be choosy.

It all boils down to costs—not financial, but energetic. Each sex bears a cost associated with reproduction, be it due to the production of gametes (sperm or eggs, bigger or smaller, more or fewer), the development of the offspring (carrying or incubating fetuses) or post-natal care (feeding the young, keeping them warm and ensuring their survival). All of these processes require energy, and, depending on the species, one of the sexes is likely to invest more into parenting than the other. But how does this investment link to choosiness?

Parental investment theory, developed by Robert Trivers in 1972, proposes that the sex investing more into parenthood is also the pickier one when it comes to mate selection, and the one investing less must compete with others to access mates. In line with this theory, when females invest more—e.g. through pregnancy, nursing, or producing large eggs—they are the ones that need to be wooed by the males' dazzling features, gifts, or courtship behavior. These displays may indicate to them how fit a potential partner is, and whether he's worth the reproductive hassle. Because females are generally the "limiting factor" when it comes to reproduction, they are the choosy sex—and there is a higher selection pressure on males to be more aggressive, impressive, or possessive (see Saiga antelope, page 64, and Guianan cock-of-the-rock, page 184).

While, in most organisms, females invest more in reproduction, there are species, such as the South American wattled jacana, *Jacana jacana*, where the males bear the brunt of parenting (the giant water bug, page 104, is another such species). Jacanas are wading birds, distinguished by their absurdly long toes. The oversized digits of Edward Scissorfeet distribute their weight over a larger surface, which allows them to walk atop lily pads, gaining them the nickname "Jesus birds." The jacanas also exhibit sex-role reversal: the females are the bigger and more aggressive sex; their wattles (the flappy skin around their beaks) are a brighter orange-red than the males', and their wing spurs—keratinous spike-like bits of weaponry that stick out of their wings—are larger. Jacana ladies are polyandrous; they have several partners on the go, and will compete for males and defend their territories. A female lays up to four eggs for each male, and then moves on to her next partner, replacing eggs as needed when they are predated. The eggs are extremely small compared to her body size and only take her about three weeks to produce; in comparison, jacana single dads look after the young for about four months.

The male is the sole caregiver—he does all the incubation and shading of the eggs, and, after they hatch, he guards the young, teaches them how to forage, and broods them under his wings. When in danger, wattled jacana chicks employ one of two escape strategies. Upon their dad's call, they jump into the water and stay there for up to half an hour, using the tip of their beak as a snorkel. Alternatively, if the threat comes from the water, they will hide under their father's wings—and, what is more unusual, he will use specialized wing bones to pick up the chicks and carry them to safety (while treating onlookers to a freakish display of long toes sticking out from underneath his feathers). The male may also feign injury to distract the predator, or occasionally call on the female to attack the intruder while he escapes with the young ones.

Apart from these sporadic defenses, the mom engages with the young extremely rarely—only when the dad dies, or when she miscalculates her laying and leaves the male to tend to a nest full of eggs *and* a clutch of mobile chicks at the same time. In such emergency circumstances, the female is able to take over all the necessary parental duties; in all other respects jacana ladies boast a rather enviable free-agent lifestyle.

Yeti crab

Kiwa tyleri

Mr. Twit, the mean and disgusting protagonist of Roald Dahl's book *The Twits*, never went truly hungry because he could always feed himself on little bits of food stranded in his bushy beard. Deep in the ocean lives nature's own version of Mr. Twit: the yeti crab, discovered a quarter of a century after Dahl's death.

Yeti crabs are a group of crustaceans in the genus *Kiwa*. They are not true crabs, however, and actually belong to a taxon called squat lobsters (a name more flattening than flattering). The genus is christened *Kiwa* after the Polynesian shellfish goddess, although one of the species, *Kiwa tyleri*, is dubbed "the Hoff crab" after the American television god David Hasselhoff. The nickname holds a clue to the yeti crabs' looks—they are covered in bristle-like setae, with *Kiwa tyleri* boasting a chest as hairy as the *Baywatch* star's. Well, technically not a chest, but the underside of the carapace. Other yeti crabs have bushy claws and legs; the hairy body parts are the equivalent of Mr. Twit's beard—that's where the food comes from.

The 5cm-long Hoff crabs are found at the East Scotia Ridge in the Southern Ocean, inhabiting hydrothermal vents—underwater fissures in places where the Earth is geologically active. There, the movement of tectonic plates causes gushes of hot, mineral-rich water, providing excellent, if somewhat extreme, living conditions for certain bacteria. These bacteria are autotrophic, meaning they produce their own food—but unlike plants and land-dwelling bacteria, which do so by converting energy from sunlight (via photosynthesis), they are chemosynthetic, and use methane or sulfur compounds as their source of fuel. The bacteria themselves then become food for other animals, establishing hydrothermal vents as veritable wildlife hotspots.

And they really are *hot* spots! The outflow at the top of active hydrothermal vents can be as hot as 380.2°C, though there are steep variations: the lower structures fluctuate at around 3–19°C, and even a few centimeters one way or the other makes a difference in heat levels. By contrast, the surrounding Southern Ocean water averages around 0°C at a depth of 2,600m. As a result, one of the hottest environments on the sea floor is surrounded by the world's coldest ocean. That's bad news for yeti crabs—as decapods, they can't handle polar temperatures, becoming inactive and paralyzed. Since their preferred temperatures are up to 24°C, they are very much restricted in where they live. Vent sites form underwater thermal islands, increasingly crowded by the mini-Hoffs, whose densities reach over 700 individuals per square meter. Competition is fierce, and the little blind crabs (who needs eyes at depths where the sun don't shine?) cling on to every feasible bit of the vent chimneys. The bigger males pick the warmest spots, while the brooding females are left with the coolest, toward the peripheries.

It's not just space that is scarce—so are resources. Sure, there are mats of bacteria growing here and there, good enough for a crab to scavenge on, but with so many keen crustaceans it is safest to grow your own meals. This is where the bristly setae come in: they provide the perfect growing grounds for bacteria. Yeti crabs of a related species, *Kiwa puravida*, inhabiting underwater methane seeps, will even slowly wave their bristled claws around to ensure optimal mineral flow for fertilizing their personal bacterial farms. Actively cultivating bacteria in your hairy claws may not sound like the most attractive solution, but it does ensure snacks are always literally on hand.

Zombie worms

Osedax spp.

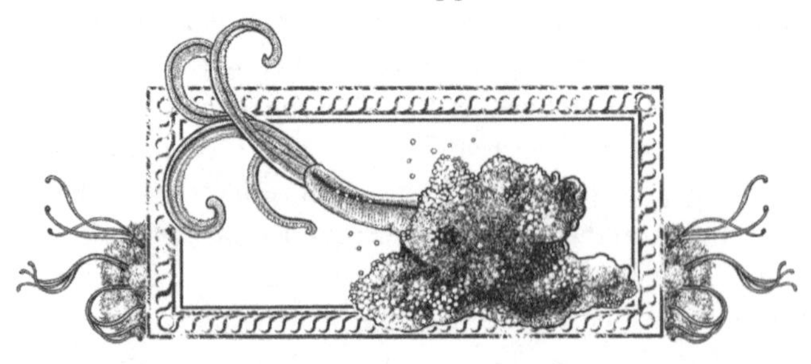

What happens when whales die? Their huge bodies fall to the ocean floor, reaching depths of several thousand meters. They arrive at the bottom relatively intact, since they sink quickly and there are no significant scavengers in the water column. Whale falls create unique, nutrient-rich ecosystems that may sustain creatures of the abyss for decades: in the cold depths, the lipid- and protein-rich whales are a rare treat. Their flesh is picked away by sharks, crabs, and hagfish (see page 108). Their skeletons are left behind, to be covered by beautiful, delicate, colorful flowers, resembling the fresh graves of beloved relatives. Except the flowers are not really flowers—they are worms. Zombie worms. And they are there to eat the bones.

Zombie worms, also known as bone-eating worms (a literal translation of their genus name, *Osedax*), are distant relatives of the more mundane earthworms. Small (just a few centimeters long), with pinkish-red "stems" and feather-duster-like "petals"—which are, in fact, respiratory palps used for breathing—they resemble plants more than animals. One species even gained the glamorous scientific name *Osedax mucofloris*, which translates as "bone-eating snot flower" worm. Like plants, these worms have roots, serving not just as an anchor but also as a way of obtaining food. This is a necessity, since, surprisingly for an animal that feeds on bones, zombie worms seem to have missed out on pretty crucial body parts: the mouth and gut. Instead, their soft root-like tissue secretes acid and enzymes to drill into the lipid-rich whale skeleton, and the dissolved nutrients are delivered to symbiotic bacteria living within the worms. These symbionts then metabolise the organic compounds, and feed the host. The acid borings are also used as shelter holes for the worms.

Bone-eating worms are pretty ubiquitous; they have been found at depths of between 30 and 3,000m, in the Atlantic and Pacific, from Sweden to California to the Antarctic. A single whale skeleton can house between half a million and a million adults. But here is the catch: those adults are all female. If you thought the worms' feeding habits were weird, wait till you find out about their sex lives . . .

All *Osedax* are sexually dimorphic: the females look very different from the males. Inside their stem-like, gelatinous tubes, the zombie worm ladies harbor harems of a hundred or so microscopic males; the older and larger the female, the bigger her harem. Every male, or should we say man-child, is hardly more than a sperm-filled larva, reliant on a yolk droplet for energy (males don't have the symbiotic bacteria to supply them with food). This polygamous relationship is a very strict-termed, no-nonsense barter: housing for sperm. The supremely fecund females produce a constant stream of larvae, and while most of the young won't make it in the ocean depths, some lucky ones will find their way to another whale fall to settle down. It is likely that the sex of the larva is determined by environmental conditions—after spawning, those that land on dead whales become females, and those landing on the females, males. Females can colonize skeletons at densities of 3–20 individuals per centimeter, and once they settle, the late arrivals become males and live inside their girlfriends.

Unfortunately, commercial whaling reduces the number of sunken carcasses, and thus removes valuable sources of nutrients for animals such as the bone-eating worms, potentially leading to a lower species richness in the depths. Fewer whale carcasses means that larvae need to cover longer distances before finding a new dwelling—a task that isn't easy anyway. And whatever their anatomical constraints might be, zombie worm females are certainly good at one thing: having sex on the graves of whales with a harem of underage males. For that unique skill, they deserve the chance to flourish.

AIR

Bees

superfamily Apoidea

Whoever decided to call The Talk a chat about "the birds and the bees" had a very strange idea about sex education. While birds can lead a generally inappropriate lifestyle (see Ducks, page 96), bees show a range of reproductive behaviors that are simply bizarre—and none of them seems very healthy by human standards.

The 20,000 or so bee species, found across seven families in the clade Anthophila, have one thing in common: haplodiploidy—a breeding system in which fertilized (diploid) eggs develop into females, while unfertilized (haploid) eggs turn into males. The reproductive female stores sperm inside her spermatheca, and can control the sex of her offspring by fertilizing the eggs at will. But the fact of the matter is: as a consequence of the haplodiploid breeding, male bees don't have daddies.

Technically, female bees also lack male role models, as there is only a single well-documented bee species, *Ceratina nigrolabiata*, where both parents look after the brood. When it comes to other bees, the situation is less than rosy. Honeybee drones (genus *Apis*) couldn't be caring fathers if they tried, as they don't live long enough to meet their offspring. Honeybees have sex on the wing, during the queen's mating flight—and the lovemaking is short and sweet, lasting from one to five seconds. As the bees copulate, the explosive ejaculation blasts the semen into the female's oviduct, sometimes with an audible pop. Unfortunately, the explosion is so powerful that the drone's endophallus (penis equivalent) is ripped off, and the male dies soon after. The honeybee's love life is best summarized with a quote from *Mean Girls*: "Don't have sex, because you will get pregnant—and die." Still, the female mates with up to twenty males, and once she is done, she is left not just full of semen, but also with the remains of torn penises inside her.

Other bees are not much better. Coercive sex is not uncommon, and female bees of many species do not have much choice when it comes to copulation, with males grasping on to them using mandibles and legs despite visible resistance. Some bees, such as the North American solitary desert bee, *Centris pallida*, are so eager to find a mate that they will literally dig her from underground, to make sure the virgin females are impregnated right after emerging from their nests. Meanwhile, the western European chocolate mining bee, *Andrena scotica*, is partial to a bit of incest, with 70 percent of the females mating with their male nest-mates. Finally, there is the issue of sex dolls: many bee males are duped by bee orchid flowers, which mimic the appearance and smell of female bees; the plants trick drones into copulating with them—so they pollinate them in the process. Each type of bee orchid caters to the tastes of a different pollinator.

In eusocial species (see Naked mole-rat, page 54), such as the European honeybee, *Apis mellifera*, characterized by cooperative brood care and a reproductive division of labor, most females practice celibacy. The workers forego sex to look after their younger siblings, produced in industrial quantities (up to 2,500 per day) by the queen. This move makes sense, as honeybees are "supersisters"—they are more closely related to each other than to their mother or potential offspring. That's because the workers inherit 50 percent of their mom's genes, but 100 percent of their dad's (who is haploid, i.e. has one copy of each gene); the sisters thus share a whopping 75 percent of genetic material with each other.

Still, some workers, such as those of the Cape honeybee subspecies, are able to birth daughters via parthenogenesis. Unfortunately, these worker clones are a nuisance to the hive: rather than working, they only lay eggs that develop into more brood-obsessed clones—and the colony fades away.

From celibacy to incest to explosive penises—perhaps the reason that people talk about "the birds and the bees" is to discuss all eventualities?

Bombardier beetles

subfamily Brachininae

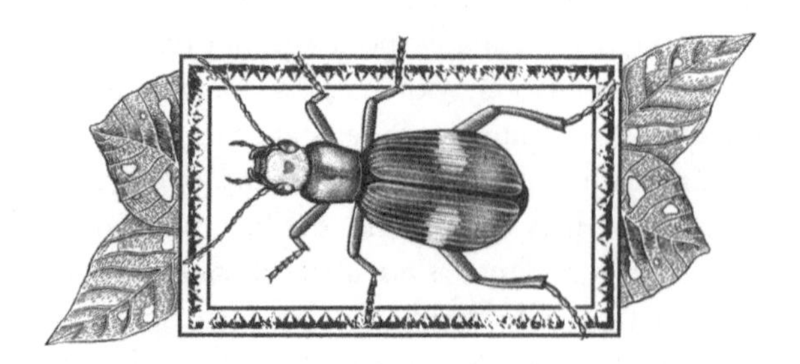

Beetles fly—but, unlike other flying insects, they can't take off very quickly. Their wings need to be unfurled from the wing covers before taking flight, a task that simply takes too long when danger is looming. Because of the delay to their escape routine, ground beetles have evolved alternative ways of protecting themselves—and the most spectacular method is used by the 500 or so species in the subfamily Brachininae: the bombardier beetles.

Bombardier beetles might not have aced the flying module in their gym lessons, but they sure paid attention in chemistry class. They produce their own chemical weapon: a toxic, scalding spray, which can be released at the enemy from one of two abdominal glands at a moment's notice. The inside of a bombardier bum is like a laboratory storage unit, with separate tanks for stockpiling different reagents. Each butt-gland has two reinforced compartments—the inner one is a large reservoir holding a solution of hydroquinones and hydrogen peroxide; the outer one, a reaction chamber, contains a mixture of enzymes—peroxidases and catalases. To deploy the artillery, the beetle squeezes some of the reservoir fluid into the reaction chamber, initiating a rapid and volatile series of events. The oxygen freed from hydrogen peroxide oxidizes hydroquinones into quinones, compounds which are highly irritant to many animals. The oxygen also acts as a propellant, triggering the substance to shoot out of the beetle's nether regions. Because the reaction is exothermic, the released energy raises the temperature of the mixture to 100°C, vaporising about a fifth of it. The detonations are accompanied by loud pops, which serve as additional deterrents for the predators. The beetles become noxious, obnoxious *and* too hot to hold.

A bombardier can direct the "nozzles" of its caustic cannon with startling accuracy—it is able to aim not only at, say, an ant sitting on its right front leg, but also at the precise segment of that leg. The sharpshooter points the spray jets in any direction—behind itself, above its head, underneath its belly, etc.—with a firing range of a few centimeters; it can fire forty-six shots before its tank is empty, and it does so at an astounding frequency of almost a thousand per second. These remarkable stats have wowed engineers, and the structure of the bombardier beetles' behinds are likely to be mimicked in improved aircraft engines.

The insect's volatile defense technique works not only prior to an attack, but also during and after. Some bombardier beetles manage to get themselves out of a seemingly fatal pickle with their cool heads and hot bums: when swallowed by a frog or toad, detonating a few explosives while inside the assailant's belly can be life-saving. The amphibian will not only spit out the foul and scorching beetle, but, if the blasts occur further down the digestive tract, it is actually able to evert its stomach to get rid of the offensive meal. To determine that this behavior is in fact prompted by beetle explosions, a research team from Kobe University listened for the pops inside the toad. Beetles who fired their shots were able to come out of their adventure unscathed, even over an hour after being devoured, and boast that they have traveled to hell and back.

Why do animals need such elaborate defense systems? Richard Dawkins and John Krebs explain the stronger selection pressures on the prey species by the "life–dinner" principle: the prey must run (or, more generally, evolve) faster than the predator, because it is running for its life—whereas the predator is only running for its dinner.

In the case of the bombardier beetle and the toad, the dinner gives the word "heartburn" a very literal meaning.

Boobies

Sula granti, Sula nebouxii

Don't snigger at the name! The word "booby" actually comes from the Spanish word *bobo*, meaning "silly," or "fool." Boobies received this rather unkind title either because of how clumsy they looked on land, or because they were tame enough to be easily caught and eaten by sailors.

There are seven species of boobies, the best known being the blue-footed ones, *Sula nebouxii*. Their bright azure tootsies play an important part in advertising their fitness and fertility: the brighter the color, the sexier the booby. Yet a bird with plain black feet, the Nazca booby (*Sula granti*), is perhaps more interesting—and certainly more ruthless—than its blue-footed cousin.

Both Nazca and blue-footed boobies can be found around the tropical west coast of the Americas, and breed on the Galápagos Islands. They lay eggs on the ground. The pair shares parental responsibilities. In both species, the females—the larger of the sexes—honk or squawk, while the males whistle. But upon closer inspection, the species' behaviors are very different.

Boobies, like other seabirds (see Laysan albatross, page 190), are generally considered monogamous, at least on paper. While Nazca boobies stay truly faithful to their mates, blue-footed boobies are partial to a bit of adultery, with over half of females engaging in extra-pair copulations. Because of such high infidelity rates, the concept of family is a bit more relaxed—stray chicks may be adopted by unrelated adults, mainly because fathers are never quite certain of their paternity. Mothers, who never have to wonder, are much more aggressive toward unrelated chicks.

While they might be decent partners, they certainly don't make great parents. In fact, if you ever think that you've failed at parenting because your kids are constantly fighting, rest assured that you are nowhere near as bad as a booby. Nazca boobies lay two eggs and incubate both; the first offspring hatches a few days before the second, and thus has a head start in terms of size and strength. When the younger sibling arrives, booby parents sit back and calmly watch the stronger chick torment the weakling. This is no mere friendly teasing—boobies take it to a whole new level of morbidity: siblicide. This phenomenon is not that uncommon among birds. Usually it is facultative—that is, it only occurs on occasions when the food is scarce and the chances of survival are small for the weaker chick. Indeed, this is what blue-footed boobies do: the older chick will tolerate the younger one, as long as there is plenty of food for both—although the moment tummies begin to rumble, the little one has to go. In Nazca boobies, however, siblicide is unconditional and obligate, meaning that killings happen every time, regardless of environmental factors. Out of two eggs, only one chick makes it to adulthood—Nazca booby youngsters tend to kill off their baby brother or sister within a week of their hatching. The evolutionary reasoning behind it is explained by the insurance-egg hypothesis, proposed by E. F. Dorward in 1962. The idea is simple: the second "marginal" egg is there in case of the first "core" egg's failure. To put it bluntly, boobies produce an heir and a spare.

Interestingly, parental intervention does make a difference when it comes to siblicide. In a study conducted by Lynn Lougheed and David Anderson, Nazca booby chicks were reared by the more placid blue-footed boobies and vice versa. Nazca chicks raised by blue-footed parents were less aggressive and the younger siblings had a higher chance of surviving. On the other hand, blue-footed youngsters raised by Nazca adults were more likely to kill each other. While blue-footed parents try to placate their young, Nazca boobies seem to egg them on!

Sadly, young Nazca boobies aren't threatened only by their own family. Around 80 percent of non-breeding Nazca adults will take an interest in unrelated chicks—and can be very brutal. While they do not explicitly kill the younglings, they may bite them, shake them, or pull their feathers. This has a lasting impact on the chicks: the more bullied a booby was in its infancy, the more bullying it will inflict as an adult. The cycle of violence continues.

California scrub-jay

Aphelocoma californica

In Catholicism, Saint Anthony of Padua is venerated as an incredibly useful saint: he is the patron of the recovery of lost items. Yet if Catholics were more like scrub-jays, Saint Anthony could take a break—these birds never lose anything!

California scrub-jays (*Aphelocoma californica*), previously known as western scrub-jays and lumped with Woodhouse's scrub-jays into one species, are medium-sized, blue-and-gray birds from North America. They are the distant cousins of magpies and Old World[1] jays, and, like other corvids, they hide away food for snacking at a later date.

Unlike those of us who regularly forget where we left our keys, western scrub-jays form integrated memories of exactly what they hid, where they hid it, and when they hid it. The ability to recall the what-where-and-when of particular storing events challenges the assumption that animals are incapable of episodic memory. What's more, scrub-jays show evidence of future planning: they can also remember how perishable a food item is so that, when retrieving the goods, they prioritize the caches with items likely to go off soon.

Scrub-jays don't focus exclusively on their own food, however—given the opportunity, they will steal supplies hoarded by others. Because they have a great memory, they are able to return to the sites where others have stored provisions and pillage them once the

---

1 While the term "Old World" has a somewhat colonial feel to it, it is—at least at the time of writing—the accepted biogeographical terminology for denoting wildlife from Afro-Eurasia. Standard usage includes phrases such as Old World jays, Old World monkeys, or Old World fruit bats (page 198).

initial owner leaves the scene. To avoid pilferers when hoarding food, scrub-jays will resort to multiple tactics. Firstly, they will try to pick hiding places in areas with few other birds. Secondly, they will hide food when nobody is looking, or when something obscures them from prying eyes. Thirdly, if a bird is present, they will try to hide food as far away as possible, and in a shaded area, to make the precise location difficult to identify. However, if the above options are not feasible, and they have to stash the morsels in the presence of an onlooker, the jays will come back as soon as the eyewitness has gone and change the location of the larder. Interestingly, changing the hiding spot is done by scrub-jays who have previously been thieving themselves; this behavior has not been noted in naive birds. They also keep track of who has been watching them, and of whom they should be wary.

When not plotting complex logistical ploys, scrub-jays, like many other birds, engage in a more pleasant pastime: spa treatments. The treatments are perhaps less calming than human spas, though, since they involve bathing in live insects. The technical term for insect spas is "anting," although other invertebrates, such as bombardier beetles (page 160) and millipedes (page 46), are also used for bird-pampering purposes.

"Passive anting" involves the scrub-jay spreading out on an ant hill, rubbing its wings and tail in it, and allowing insects to crawl over the feathers. During "active anting," the bird picks up insects, crushes them with its bill and rubs them through its plumage. The exact purpose is unclear; however, it is likely to help feather maintenance and combat parasites or bacterial and fungal infections using the ants' formic acid (or other substances released by insects). Some birds choose to eat the ants that have been "emptied" of their unpalatable toxins. Saint Ant would probably not approve.

Caribbean reef squid

Sepioteuthis sepioidea

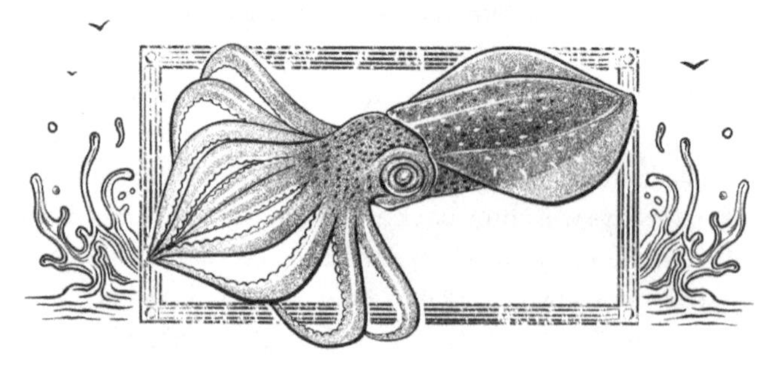

With its "live fast die young" attitude, the Caribbean reef squid, *Sepioteuthis sepioidea*, probably wins the title of the rock star of the underwater world. A resident of warm waters from Bermuda to Brazil, the species is on the smallish side (certainly compared to the record-holding 13m giant squid), with a mantle, or the "hood," containing most of the vital organs measuring only about 20cm. Still, size does not matter if you have style, charisma, and ultimate coolness.

These molluscs don't ever sit still. Since they lack a swim bladder, they need to be in constant motion—otherwise they sink. But while the Caribbean reef squid might not drive a motorcycle like James Dean, it can do something way better. When feeling lazy, threatened, or just extra-hip, rather than swimming, this squid will . . . fly. Why? Because it can—and because flying is actually likely to be less energy-expensive.

Squids have two modes of locomotion: they can propel themselves by releasing a jet of water from the funnel placed between their arms, or they can paddle around using two rounded fins on both sides of their torpedo-shaped mantle. The jet-wielding funnel can be aimed forward or backward, thus allowing the animal to scoot along in any direction. But, as anyone wading through water will know, its density can get in the way of efficient movement. To overcome this, squids, aided by their rocket-like funnel jets, will jump some 2m above the surface, spread their wing-like fins, and fly at least fifty times their body length. When they intend to brake, they flare out their arms and land back in the water. This behavior of the Caribbean reef squids was first observed in 2001 by Silvia Maciá—and it led to the identification of similar abilities in six other squid species. Some are even likely to spread a sheet of mucus between their arms to enhance their flight capacity.

Unlike other cephalopods, the Caribbean reef squid is not too cool to school—it is a gregarious species. Consequently, its social communication is likely to be much better than that of solitary molluscs. Being active in well-lit, shallow waters means that it's reliant on looks when it comes to social interactions. Squid schools resemble college parties, with "scramble competition," where everyone tries to find a mate for themselves. Much like the freshmen "traffic light parties," where participants dress in red, amber, or green to show if they are single or not, squids will signal their availability to conspecifics using visual cues. Except the under-the-sea raves don't use colors, since squids are color-blind—they use patterns.

With the help of chromatophores, pigment-containing cells that contract or expand as needed, Caribbean reef squids can don a total of sixteen different outfits—including flicker, zebra, saddle, or stripe. These skin patterns indicate the animals' eagerness to mate, but also may serve as a warning to competitors. And when threatened, squids will put on a deimatic display (see Red-eyed tree frog, page 60), and flash two dark "eyes" on their mantle. Furthermore, they squirt ink to disorientate the opponent—which is also a signal for the other squids to jet off.

Squids are proponents of consensual free love—both males and females mate with multiple partners, though females tend to lose steam and enthusiasm after several matings. After flashing appropriate patterns and engaging in a bit of chasing, the couple will do a mutual rocking swim, which is followed by the male depositing his spermatophores at the bases of the female's arms. The spermatophores are not transferred under the female's mantle without her consent: for squids, no means no!

Unfortunately, producing young puts an end to the squids' carefree life. The species is semelparous (see Antechinuses, page 14), which means that after the males mate and the females lay eggs, they die, having lived for less than a year. They never meet their offspring, who, without older and wiser role models, are doomed to repeat their parents' rock-and-roll lifestyles.

Chatham Island black robin

Petroica traversi

In the beginning, there was an archipelago. More precisely, Chatham Islands, about 800km east of New Zealand's South Island. It was a Garden of Eden for birds, inhabited by many endemics—species not found anywhere else. Unfortunately, the arrival of people, especially European settlers, and the cats and rats associated with them, spelled doomsday for local avifauna. Unaccustomed to mammalian predators, having evolved in their absence, birds such as the Chatham raven, Chatham fernbird, and Chatham rail went extinct. The Chatham Island black robin, *Petroica traversi*, a sparrow-sized black fluff-ball, was thought to have shared their fate—until, in 1938, a small population was discovered on the island of Little Mangere.

Little Mangere, no more than a 200m-high rock stack, with some 5 hectares of rapidly degrading scrub and forest on top, proved less of a paradise. By 1973 it housed only eighteen black robins, and six years later that number had declined to a total of seven. These seven survivors were relocated to the neighboring predator-free Mangere Island, where newly planted forest provided a better habitat. Still, they did not fare well, and by 1980 the entire black robin population consisted of five individuals—including one viable female, named Old Blue after the color of her leg band.

At this point, a guardian angel appeared: Don Merton from New Zealand Wildlife Services, who took a more interventionist conservation approach. Black robins are a slow-breeding species; they take a couple of years to mature sexually, live to be about six, and lay only two eggs per clutch. Yet when a clutch is lost, they are able to produce another one—a feature that Merton exploited. Remembering an old childhood prank, where he convinced his grandmother's canary to look after goldfinch eggs, he started a cross-fostering program with the robins—the first ever in an endangered passerine bird. Initially, robin eggs were fostered by Chatham Island warblers (*Gerygone albofrontata*). However, unable to raise the young past the age of ten days, the warblers were replaced with Chatham tomtits (*Petroica macrocephala chathamensis*), who made better adoptive parents. With the tomtits doing the incubating, the robins were left to focus on the laying, thus boosting the output of precious eggs.

Between 1979 and 1981, Old Blue and her mate, Old Yellow, were the only successful breeding pair, saving the species from extinction and becoming the Adam and Eve of all black robins currently in existence. Miraculously, Old Blue, a veritable Methuselah, lived to be at least thirteen, while maintaining steady egg production to the very end and raising a total of eleven chicks.

In 1984 an observation was made: some females laid their eggs not in the middle of the nest, but on the rim, where they could not be incubated properly. At the early stages of the recovery program, every egg was invaluable; thus, Wildlife Services staff repositioned the eggs in the center of the nest, to allow proper development—and continued to do so until 1989. By that time, over half of black robin females laid rim eggs, suggesting that the behavior was hereditary. The well-meaning staff had unknowingly promoted a maladaptive trait—i.e. one reducing the offspring's survival chances—by not allowing natural selection against it. Fortunately, in 1990, with a more hands-off conservation approach, the egg repositioning interventions stopped—and subsequently the proportion of rim-laying females dropped to around 20 percent.

The latest black robin census clocks their number at around 250. Even though the population is severely inbred, leaving them vulnerable to diseases, the species numbers are increasing; the bird is currently listed as endangered. Birds appear surprisingly resilient to genetic bottlenecks, with a number of other species brought back from the brink, such as the pink pigeon (down to ten individuals), Laysan duck (seven individuals), or the record-holding Mauritius kestrel (four individuals). Despite the shockingly low numbers, these species all are fruitful and multiply. And it is good.

Common pigeon

Columba livia

In 1936, Jan Wedel, a Polish chocolatier, came up with a name for his company's latest product: *Ptasie mleczko*—"Bird's milk"—a phrase denoting an unobtainable delicacy. However, bird's milk itself is not as unattainable as lore implies; to get some, Wedel could have simply scattered grain in front of his shop, since the mythical delight can be obtained from the humble pigeon. Thankfully, Wedel's sweet is a chocolate-covered vanilla delight, rather than a concoction of fluid-filled cells shed from the lining of a pigeon's crop.

All pigeons and doves (as well as flamingos and emperor penguins) feed their young "crop milk"—a semi-solid, fatty, protein-rich substance regurgitated by both parents from a pouch in their gullet. Baby pigeons, called squabs, receive this unappetising diet for the first week of their lives, after which they are gradually introduced to grains, still soaked in their parents' crop juices. This happens because, upon hatching, squabs are unable to digest adult food—they are also blind, covered in wispy fluff, and have oversized beaks; literally no part of them looks presentable.

In the wild, squabs hatch in holes in cliffs. But what is "wild" has become very fluid for pigeons—much like the birds themselves. Common pigeons started off as rock doves, *Columba livia*, native to the Mediterranean and western Asia. They became domesticated at least 5,000 years ago (forming the subspecies *Columba livia domestica*), but then many chose freedom again. These feral pigeons are now widespread, ranging far beyond the original rock dove distribution. And because man-made structures are perfect imitations of rocky cliffs, it is no surprise that pigeons do so well in urban areas.

Domesticated pigeons have been bred for food (what the milk-fed squabs lack in looks, they make up for in taste), companionship, status and, perhaps most importantly, communication. Their incredible navigational skills have been used for sending messages in times of peace—for instance by Paul Reuter, the news-reporting pioneer—and of war, be it by Julius Caesar, Genghis Khan, Second World War soldiers, or the so-called Islamic State.

The premise is simple: take a pigeon to a new place, release it, and it will—almost magically—find its way back home. They can do so quickly (at speeds of 97km/h) and over long distances (up to 1,800km), though they only fly in one direction (to the home loft), so a reply has to be sent via a different bird. During the Franco-Prussian War (1870–71), pigeons were shipped out of besieged Paris in hot-air balloons and, once safe, could be sent back with messages to the Parisians. A First World War hero pigeon, Cher Ami, saved 194 American soldiers from friendly fire by delivering a message despite sustaining serious wounds.

The underpinning mechanisms of pigeon navigation still remain somewhat of a mystery. Studies with GPS loggers show that on familiar routes, birds use visual cues such as roads and other landmarks to orientate themselves. But the real challenge is finding home when released in a completely unfamiliar location, sometimes hundreds of kilometers away.

One of the proposed explanations suggests that the birds can navigate via detecting the Earth's magnetic field. This is tricky—while birds do have an internal magnetic compass for determining direction, it can only be useful if they know where they are in relation to their loft. It is therefore more likely that pigeons smell their way home. Combining the knowledge of scents around their loft with the direction of the wind gives them a good idea of which way they should fly. In 1971, Italian zoologist Floriano Papi demonstrated that pigeons deprived of a sense of smell were not able to locate their lofts; subsequent studies confirmed this finding. In contrast, magnetic manipulations did not affect pigeons' homing ability.

While some populations of feral pigeons are as old as domestication itself, various feral flocks have shown a closer genetic resemblance to homing breeds than any other type of pigeon. This suggests that wayward homers might contribute substantially to the populations we see around our city streets—perhaps making urban pigeons more veteran than vermin.

Common potoo

Nyctibius griseus

"Branch out!" they say. Well, young potoos take this advice to heart—the species' main goal in life is pretending to be a broken piece of tree.

Potoos—related to nightjars and Australian frogmouths—are a family of seven neotropical bird species, with the common potoo, *Nyctibius griseus*, being the most widespread. These birds inhabit areas from Mexico to Argentina—but unlike the tropical parrots or toucans, they are not brightly colored and flamboyant. The potoos' plumage is incredibly drab, and for a good reason: it allows them to seamlessly blend in with the trees they pretend to be a part of.

In addition to the grayish-brown feathers, the pigeon-sized birds mainly consist of a mouth and unbelievably googly eyes, very much like something from *The Muppet Show*. The common potoo's peepers are modeled after the Gruffalo: protruding, bright yellow, with black pupils. By contrast, the great potoo, *Nyctibius grandis*, has incredibly creepy jet-black eyes that seem to drill into the soul of whoever it is they are looking at. These species are nocturnal, and use their humongous headlights to spy insects (particularly beetles, moths and grasshoppers), which are swallowed whole by the ridiculously large potoo mouths. To hunt, the birds make quick dashes from their favorite branch and snatch prey mid-flight; they are not big on walking, and so do not try to pick insects from the ground.

While their enormous eyes are very useful for their nocturnal hunting escapades, they are a dead giveaway to any predators during the day. However, the potoo's defense mechanism is based around a brazen attitude and nerves of steel: it spends its entire days in broad daylight, motionless, disguised as a broken log. When danger approaches, the bird closes its eyes and mouth, and slowly stretches its head up, to make itself even more branch-like. The threat is constantly observed, either through a very slight squint, or, if the eyes are fully closed, through two small notches in the eyelids. The potoo keeps its eyes closed so as not to give away its position; it might shift ever so slightly to monitor what the threat is up to but there are no sudden movements, everything is nice and steady. While it might be playing a constant game of chicken and not moving until the last moment, if the predator ventures too close the potoo will either fly off, or abruptly open its freaky eyes and beak to scare away the opponent.

Life doesn't change much for the potoos when it is time to start a family. They don't bother with a nest—the female picks her favorite broken branch, stump, or fence post, ideally with a bit of a depression or knothole on top, and lays a single egg. At night, the monogamous birds take turns incubating the egg; during the day, it is mainly the male's job. It does not seem like much of a sacrifice—after all, he is spending his days sitting still anyway; he might as well do it over an egg. By the time the youngster hatches, it is ready for the role it was born to play: that of a broken branch. The chick remains stationary, initially huddled in its parents' feathers, and later, once it outgrows the cozy hideaway, next to Mom or Dad, or on its own. The youngster's camouflage differs from that of the adults: the whitish, fluffy plumage resembles a fungus-infected tree. Still, it does the job. After fifty or so days of standing motionless with its parents, the young potoo is ready to start standing motionlessly solo. From a low-maintenance childhood, to a smooth period of adolescence, to a chilled-out adulthood—these birds may not have accumulated a store of exciting anecdotes in their lives, but they know how to elevate monotony to an art form.

Common swift

Apus apus

For centuries, Europeans have asked themselves what happens to swifts when they are not around. The birds fly over European heads between May and September, and then disappear—where to? Aristotle suggested that, along with swallows and martins, they hibernate; the view that swifts' hibernacula (winter shelters) are found in mud at the bottoms of pools persisted for almost two millennia. In the late eighteenth century, Gilbert White, regarded as England's first ecologist, investigated the matter by having laborers dig up likely overwintering spots; since he didn't have any luck finding hibernating birds, he instead leaned towards the idea that they migrate.

It turns out that White was right. Common swifts, *Apus apus*, are only European for their breeding season; the remainder of the time they are African. These charcoal-gray birds are masters of the sky. In flight, they are shaped like a boomerang, with wings longer than the body, a short, aerodynamic tail and tiny feet used for clinging (the scientific name, *apus*, "without feet," reflects the ancient belief that the swift was a type of footless swallow—though they are actually more closely related to hummingbirds).

Swifts are the most aerial of birds; from the moment they leave their nesting sites, they can spend ten months on the wing, non-stop. As insectivores, they feed on anything they can catch in flight, and their beaks, though very short, have a very wide gape, allowing them to gulp morsels 2–10mm long. They are not picky, and have been found to eat at least 312 different prey species in the UK. Swifts also drink in flight, collect nest materials in flight and, uniquely, mate in flight. They even, presumably, sleep in flight—most likely on ascents to high altitudes—though the details of these aerial naps are still a mystery.

The only time swifts land is to raise young; since they are faithful to their nesting site, it is possible to observe them year on year. In the nineteenth century, Edward Jenner, of vaccine fame, was one of the first to study this aspect of their behavior.

As bird ringing had not yet become established, he marked the swifts by the rather unpleasant method of cutting off individual toes (at least he realized they have feet), and noticed that particular birds returned to the same nest each spring, fixing it as needed. In 1948, David and Elizabeth Lack started a study of a swift colony breeding at the Natural History Museum in Oxford, which continues to this day—one of the longest continuous studies of a single bird species in the world.

Nowadays, thanks to lightweight geolocators, which determine location based on ambient light levels, we know what swifts are up to on their journeys. To reach their overwintering site in sub-Saharan Africa, swifts take a longer route with multiple fueling-up stopovers (which does not mean they stop flying, merely that they fly around in one insect-rich location); on average, they travel 500km a day. In spring, they choose a more direct route that is around 2,000km shorter, and they cover it much faster, at over 800km a day, assisted by a 20 percent higher tailwind. Since swifts can reach an age of more than eighteen years, record holders are likely to have covered over 6 million kilometers in their lifetime—eight times the distance to the moon and back.

To prepare for such taxing journeys, swifts behave like professional athletes from an early age. While in the nest, they practice doing "push-ups" on the tips of their wings: they lift their body clear of the floor for a few seconds at a time. They also watch their weight—the push-up allows them to gauge if they are not too heavy relative to their wing length, and, if they are, they fast until they reach their target weight. In the avian Olympics, they are undisputed champions of non-stop flight.

Common vampire bat

Desmodus rotundus

This is a story of blood, camaraderie, and self-sacrifice. But it's not a tale from the trenches of the First World War, or some noble peasants' revolution—it's about vampire bats.

There are three vampire bat species, all native to Latin America, and they are the world's only haematophagous mammals, meaning they feed exclusively on blood. They are smallish (ca. 9cm long, with a wingspan about double that) and long-lived—some females reaching the ripe age of thirty in captivity. The common vampire bat, *Desmodus rotundus*, feeds on mammals, whereas the other two species prefer birds. While the main mammalian prey is livestock, bats do sometimes resort to biting humans, especially those who sleep outside, raising public health issues such as the transfer of rabies and other diseases.

Dracula's children are perfectly adapted to their sanguineous diet. Common vampire bats have the fewest teeth of any bat species, but their eighteen choppers are razor-sharp. They use echolocation (see Moths, page 194) to orientate themselves over long distances, and then identify their preferred prey individually by listening out for breathing patterns of sleeping animals. Once on the host, they sense heat with specialized proteins in their noses to select bite sites where warm blood is near the skin. After making the incision, vampire bats lap up the trickling blood using their specialized grooved tongues; clotting is prevented by anticoagulants in their saliva. The name of that anticoagulant? Draculin. Yes, scientists are geeks.

Each liquid meal lasts between ten minutes and an hour, and may increase a bat's weight by up to three times. To avoid flying with an unnecessary load, bats pee out excess water within a few minutes of starting their dinner. Unlike other bats, vampires are very good at walking and running on all fours—at speeds of up to 1.2m/s—and they even jump. This skill comes in handy when searching for the perfect dining spot on a large-bodied animal.

Yet vampire bats are not just bloodthirsty gore machines—there is a gentler and warmer side to them. They live in colonies of up to several hundred individuals; within them, groups of ten to twenty females will form close-knit associations. They recognize each other's calls, and spend time grooming each other. They also dine together.

If a bat fails to get supper, it's in trouble—it can only survive for seventy hours before starving to death. Thankfully, pals from the same colony have it covered: the more successful foragers will share a blood meal (via the not-very-appetizing means of regurgitation) with the famished friend. Vampires clearly have strict rules of fair play: a bat is much more likely to receive food if it has donated some previously. And help goes beyond family ties—female bats nourish not only their offspring and other relatives, but also unrelated adult roost-mates; reciprocal regurgitation is more important than relatedness or mutual grooming behavior.

In the wild, it is usually females who share food, mainly because males don't form stable bonds as often. A hungry bat is usually fed by several others, building a "support network" of social bonds. Also, donors are eager and willing: they are more likely to initiate the food donations than the recipients—and, shockingly, sometimes offers from keen benefactors are rejected by beneficiaries, indicating some patron preference (always ask yourself: whose regurgitated blood would you fancy for lunch?).

Bat solidarity extends beyond divvying dinner. A common vampire bat female was reported raising the orphaned offspring of a dead colony-mate—a closely associated, though unrelated, female. When the friend's health started to decline, this female spent more time regurgitating food and grooming both her and her pup; eventually, she began nursing the baby after the mother died. It was an impressive fostering commitment of at least nine months, as the weaning takes three times longer than for other bat species in the same family. Non-kin adoption has been documented in other vampire females, too. Far from being selfish bloodsuckers, these creatures demonstrate a social conscience that puts most humans to shame.

Dragonflies

suborder Epiprocta

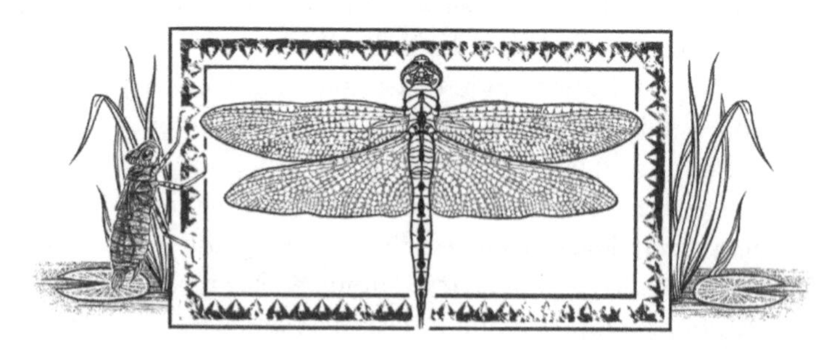

Dragonflies can fly like nobody's business. They can fly backward and forward, up and down, and side to side; they can hover, change direction at the blink of an eye, and even fly in tandem when mating. Out of the 3,000 dragonfly species (suborder Epiprocta), only a few dozen migrate, but those that do, do so expertly. The globe skimmer, *Pantala flavescens*, boasts the longest migration route of all insects: a multigenerational journey of 18,000km, with individuals traveling some 6,000km, including a non-stop transoceanic trip from northern India to Somalia. They fly over the open ocean and across the Himalayas, at altitudes of 6,300m. As if that wasn't enough, the dragonflies' maneuverability is impeccable, as their four wings can move independently of each other. With such extraordinary flight skills, these insects are exceedingly successful predators, catching up to 95 percent of the prey they pursue.

Yet, unexpectedly, these dazzling, carnivorous stunt helicopters spend most of their lives . . . underwater. Together with damselflies, dragonflies belong to the order Odonata. While some insect orders, such as butterflies and beetles, are holometabolous—that is, they undergo complete metamorphosis: egg–larva–pupa–adult—dragonflies are hemimetabolous, meaning that their metamorphosis only includes three stages: egg, nymph, and adult. Like the adult form, dragonfly nymphs are predatory—but, unlike the adults, they are aquatic.

Even though the nymphs—also known as naiads—are named after the water goddesses of ancient Greece, they are not dainty and delicate like their namesakes. Instead, they are more like water ghouls: they are voracious hunters, who not only feed on aquatic larvae of other insects (including fellow dragonflies), but also capture tadpoles and the occasional small fish. As they feed, they grow—and as they grow, they moult. Some do this as many as seventeen times before reaching their adult form; they can live between a few months and a few years in their underwater incarnation. Because of their large size (with some species reaching almost 10cm), naiads are important aquatic predators, particularly in ponds that cannot sustain fish, for instance due to drying out. Meanwhile, they themselves fall prey to waterfowl or larger fish.

To enable their underwater existence, young dragonflies have one particularly useful, if unexpected, feature—their butts. Nymph butts are the Swiss Army knives of derrieres—multifunctional tools that equip the bearer for every eventuality. More precisely, there are four main backside functionalities that a baby dragonfly makes use of. The first one is the regular, unremarkable bum function: egestion of waste. The second, much like in the Mary River turtle (page 118) or the sea cucumber (page 138), is breathing—through internal gills situated in the rectum. The third function is a nifty escape response. When a sphincter muscle in the head area is closed and the abdomen contracts, the nymph backside fires a stream of water at up to 50cm/s, making the animal jet-propelled—handy for a quick getaway. The final role of the rear end is something that must have inspired the creators of the 1986 movie *Aliens*—it's an extendable jaw. By closing another sphincter, this time in its anus, the nymph uses hydraulic pressure to shoot out an extensible, armored lip (labium). The hinged labium, armed with spikes and hooks, is folded underneath the body while at rest, and sprung out at great speed during a hunt. These butt-powered predators sure put the "ass" into "badass."

The uncompromising attitude of the dragonflies lasts into adulthood. In what must be the harshest let-down ever, female moorland hawkers (*Aeshna juncea*) who don't fancy a male will avoid interactions with him by faking their own death. When approached by a male, females, who were happily flying around up to that point, will abruptly crash to the ground and remain there motionless, feigning death, until the unwanted suitor leaves. It's a slightly extreme tactic to try next time you find yourself on a bad date, but no doubt brutally effective.

Emerald cockroach wasp

Ampulex compressa

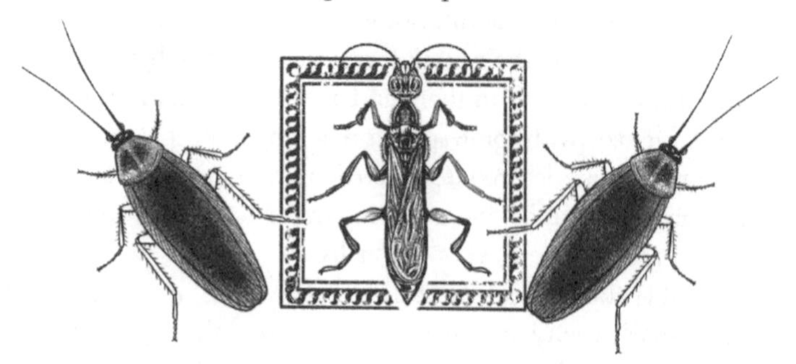

Elegant, classy, shimmering—it is no surprise that the emerald cockroach wasp, *Ampulex compressa*, is also called the jewel wasp. The species is found in tropical areas of Asia, Africa, the Pacific islands, and Brazil; with large eyes, a metallic green-blue body and red middle and hind legs, this 2cm-long insect does look as if it has just emerged from a Fabergé egg. Yet, after getting to know the emerald cockroach wasp females, it is unlikely that anyone would call them "precious"—they are the arthropod equivalent of Nurse Ratched.

Wasps are probably some of the most sophisticatedly devious of all insects—see White butterfly parasite wasp, page 212—and the jewel wasp embodies the perfect combination of beauty and cruelty. Like other members of its family, this species is a cockroach-hunting, parasitic wasp; the females use their prey as a source of food for their larvae. The jewel wasp does not, however, limit itself to simply killing a cockroach; its modus operandi is a picture of refined barbarism.

Before serving as a living larder for young wasps, the cockroach needs to be subdued and taken into a nest. Few animals would make such a sacrifice voluntarily (some cockroaches try to deter the attacker by kicking and biting), which poses a logistical problem for the jewel wasp mom-to-be, as the cockroach is as big as, or bigger than, herself. Carrying a thrashing and resisting insect this size is not an option; it is easier to make the six-legged sacrificial lamb come willingly.

The subjugation happens in two steps. First, the cockroach is stung in the thorax, which induces a temporary paralysis of the front legs, lasting two to three minutes. This restraint is necessary for the second step, when the wasp performs brain surgery on the roach. She does so by delivering a precise sting in its head ganglia. The accuracy of her aim—redolent of the most advanced drug delivery systems—turns the patient into a zombie. The cockroach is not paralyzed, but stupefied. It spends the first half hour after the operation grooming itself; subsequently, the wasp's venom induces hypokinesia, or sluggishness. Victims are still able to groom, flip over when placed on their back, or fly—but their escape response is very much impaired. What's more, the sting modifies the roaches' metabolism, and, as a result, they consume less oxygen, lose less water and generally survive longer—excellent preparation for being a living food-storage unit.

Once the prey is thus subdued, the jewel wasp cuts off one of its antennae with her mandibles and takes a swig of its haemolymph—the insect equivalent of blood. She does it either to check the roach's suitability as a host for her babies, or, more likely, to get some extra protein to boost her own egg production. She then grabs the newly docile cockroach by an antenna and leads it, like a dog on a leash, to a nest, where she lays one or two eggs between its legs. The roach doesn't protest, but lets itself be walled alive in the nesting chamber, as the jewel wasp blocks the entrance to the nest with small pebbles.

About five days after the wasp egg hatches, the larva will chew its way into the cockroach's body through the thin cuticle on the legs. It will feed on the internal organs of the host—while it's still alive, although apathetic—until, after a further three days, it pupates. Five weeks later, an adult wasp will emerge from the carcass of the finally deceased cockroach.

Under laboratory conditions, the emerald cockroach wasp parasitises a new roach every other day for a total of two months. In the wild, this number is probably lower, as it can be difficult to easily find a target for her next baby's morbid nursery.

Flying fish

family Exocoetidae

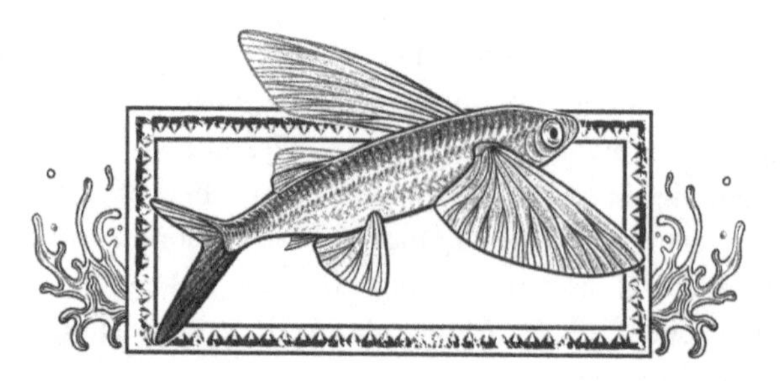

Flying is a great way to travel—if you have the right kind of wings. And even if you don't, gliding is still a good option: simply climb to the top of the highest tree and leap, making sure to mitigate the crashes while you are airborne. Most gliding vertebrate species—including snakes (see page 202), geckos, flying squirrels or frogs—come from Southeast Asia; this is probably due to the combination of forests with some of the tallest trees, and the relatively few lianas between them to travel on. Rather than slowly descending down a 60m tree and ascending the next one, it is easier and quicker to just jump across. Yet some gliders have no need of trees; they fire themselves from the depths of the sea to ascend to the skies.

Flying fish are such animals. Widespread in warm oceanic waters around the world, and measuring between 15 and 50cm, they comprise sixty or so species in the family Exocoetidae. The family is named from the Greek *ex*, "outside," and *coitos*, "bed," due to the fact that, at least according to Pliny the Elder's *Natural History*, they come ashore to sleep. In reality, they don't—but the name stuck.

For flying fish, these aeronautical adventures are not about spending a cheeky night out on land—they're a means of escaping marine predators, such as tuna, swordfish, or marlin. To take off, the fish, like aeroplanes, rely on speed. Before lift-off, flying fish (at that point still swimming fish) approach the surface at a rate of 20–30 body lengths per second, with their wing-like fins furled at the sides for a more streamlined shape. They then leap out of the ocean, unfold their wings, and enter the so-called "taxiing flight," during which they furiously beat the water with their tails at 50–70 strokes per second. After thirty or so tail strokes, they take off. During free flight, with their wings stretched and their tails high and still, they look like a fleet of fighter jets. Upon descent, the tail is lowered, and the fish might taxi before another take-off, or submerge. The latter might be a better option in the presence of airborne predators, such as frigate birds.

Like monoplanes and biplanes, flying fish come in two varieties: two-wingers and four-wingers (though the wings of the latter are not stacked). The former glide thanks to their enlarged pectoral fins, while the latter use both the pectoral and pelvic fins as lifting surfaces, resulting in longer flights. Interestingly, research by Jacob Daane revealed that overgrown fins are caused by genes influencing the flow of potassium into cells, which in turn affects embryonic development and tissue regeneration; lab-bred zebra fish (a species with normal-sized fins) with mutations within these genes developed wings akin to those of flying fish.

As if that was not enough, all flying fish species have an asymmetrical, vertically forked tail, with the longer, stiff prong at the bottom acting as a rudder. They know where they are going thanks to their eyes with flat corneas; this specialized shape allows them to focus well both underwater and in the air.

Kitted out with such adaptations, these Red Barons of the ocean achieve very impressive stats. They glide up to 8m above the surface, with airspeeds of 15–20m/s. When flying, the fish maximize for distance rather than time out of the water; they do so remarkably well, with record glides reaching 400m, twice the length of land-based, arboreal gliders. In terms of gliding performance, fish wings are comparable to those of hawks, petrels, and wood ducks; the wing loading—how much mass is carried by total wing surface area—of the largest species is similar to that of pelicans and cormorants.

With such aeronautical accomplishments, flying fish could give swifts and albatrosses a run for their money—if only they breathed air!

Guianan cock-of-the-rock

Rupicola rupicola

"Step this way, step this way, ladies, for the best show in northern South America!" Like feathered circus barkers, a group of male birds loudly solicit passing females. This is their moment, their chance to shine; these males do very little in life aside from readying themselves for the performance.

The species they belong to is the best proof that ornithologists should sometimes take a break from naming animals—it's the Guianan cock-of-the-rock, *Rupicola rupicola.* The males are bright orange, crested, and unmistakable—at least not for any other bird; they could, however, be mistaken for a chunky carrot with a pizza cutter for a head. The gray-brown females sport a much smaller crest, and are certainly more average-looking.

In line with their appearance, female cocks-of-the-rock have a down-to-earth, no-nonsense approach to life. They build solid nests out of mud and vegetable fibers in rocky areas (hence the species name), and repair them over the years. They lay one or two eggs, and raise their young as single mothers. They might bump into the males when foraging on the same fruit trees, but that's about it, most of the time.

Males, meanwhile, devote their lives to making an impression. Cocks-of-the-rock are a lekking species, which means that they engage in competitive displays to seduce females. A lek, from the Swedish word meaning "play" or "game," is an area where such displays take place. Each male secures himself a court—a patch of forest floor 1m across, cleared of all leaf litter—and a nearby perch. Cock-of-the-rock leks can contain fifty or so such courts, and the central ones are of the highest value, a bit like the merry-go-round in the middle of a village funfair. When a female flies over the site, she is greeted by loud vocalizations from the males, and treated to elaborate wing-beating courtship dances on the courts. After a few minutes of this pandemonium, if she is interested, the female will perch some 2–6m above the lek, to have a gander without committing to any of the territories—at which point the males crouch on the ground and adopt their display positions. Here, stillness is crucial. The male flattens himself on the ground, completely motionless, with his back—embellished with a fluffy cushion of orange feathers—facing the female. If she likes what she sees, she will land in his court for a closer inspection, during which he puffs up even more, and, with the crested head to the ground, tries to look his pizza-cutteriest.

Eventually, the hen makes her choice, letting the lucky male know with a tap that she is ready to mate. As that happens, there is a round of heckling from all the other males—"copulation alert calls," or loud screams, in one last attempt to redirect the female's attention. If the hen is not put off by this, the mating goes ahead. It lasts a glorious 10–15 seconds, to the accompaniment of shrill alert calls from the neighboring males (presumably along the lines of "They're doing it!"). Afterwards the female generally loses interest in cocks until the next breeding attempt. Because the species is polygynous, some males secure multiple matings, while others (sometimes more than half of the lek) don't get to mate at all.

A centralized lekking ground makes the choosing energy-efficient for the female—she does not need to go around inspecting individual territories; a quick scan is enough to assess the goods. The sole reason for her visiting a lek is a hook-up; the grounds contain no extra resources for food or building a nest. The males need to make themselves conspicuous and, in the dark forest, the fluorescent-orange birds give the impression of flames. However, they also attract the attention of predators—birds of prey, wild cats, or snakes. Since just a few successful individuals are enough to keep the species going, the risk-taking males are the more expendable sex. On the plus side, leks provide safety in numbers: many displaying males equals many anti-predatory lookouts. There's a practical benefit to this unique orange-light district.

Hummingbirds

family Trochilidae

Feed like a butterfly and fly like a bee—hummingbirds might have twisted Muhammad Ali's famous quote to suit their needs, but one thing is certain: the little bird can claim to be as tough as the famous boxer. The 360 or so species in the family Trochilidae seem to share a single aspiration—to be like an insect (but better). And, as they follow their grand plan, hummingbirds break records and defy expectations.

Found across the Americas—from Alaska to Tierra del Fuego—these exquisitely plumaged birds are split into subfamilies with names as delightful as topazes, emeralds, brilliants, coquettes, and mountain gems. The smallest species, the bee hummingbird (which clocks in at 2g, the weight of two thumbtacks), is also the world's tiniest bird. The giant hummingbird, weighing just ten times more, is the largest of the lot. After all, a wannabe insect cannot be too big.

Hummingbirds take feeding like a butterfly very seriously—they are specialized nectarivores, with a range of adaptations to prove it. While birds as a group lost their ability to taste sugars, hummingbirds have secondarily evolved a sweet taste perception—and a preference for the most sugary food on offer. Their forked, grooved tongues pump and lap up nectar at a rate of 15–20 licks a second. Such efficiency is crucial for meeting the bird's daily calorific intake—hummingbirds feeding on dilute sugar nectars consume over five times their body mass per day. When not gorging on nectar, they snack on tree sap and fruit, catch invertebrates for a bit of protein, and even eat soil to obtain minerals.

As so-called trap-line feeders, hummingbirds visit the same plants on a regular basis (like a trapper checking lines of traps), and carry pollen between them. The birds' pollinating services are rewarded with nectar at the bottom of the elongated flowers they target. These partnerships may lead to coevolutionary escalation: longer-tubed flowers require longer-beaked hummingbirds—culminating in almost ridiculous adaptations. The sword-billed hummingbird, *Ensifera ensifera*, a record holder for relative beak length among birds, is the only bird with a bill longer than its body. Yet despite its competitive advantage as a foraging tool, this épée of a beak is so cumbersome to carry that, when perching, the bird points it upward to reduce muscle strain. It's also useless for preening—swordbills resort to cleaning themselves with their feet.

Having an insect-like proboscis is not enough; hummingbirds also fly like insects. Like bees, the birds approach flowers by hovering, and withdraw by flying backward—constantly beating their wings, up to eighty times per second. This extraordinary aeronautical achievement requires a host of physiological adjustments: hummingbirds boast humongous pecs (taking up over a quarter of their body mass) and the biggest avian hearts, relatively speaking. To cover the energetic expense of hovering, they have the highest metabolic rates of all vertebrates. Their legs are reduced to the bare minimum, and they have the fewest total feathers of all birds—sometimes under a thousand—to shed unnecessary weight. And if they are not flying, hummingbirds can save energy by going into torpor, a hibernation-like state where body temperature drops from 40°C to 18°C, and the heart slows from over 1,200 beats per minute to under 100.

As if this was not remarkable enough, hummingbirds up their flying game when they go courting. Male Anna's hummingbirds (*Calypte anna*) perform aerial dives at a record-breaking speed of 385 body lengths per second—twice that of a diving peregrine falcon—making them the fastest vertebrates, relative to body size. The tiny birds endure centripetal accelerations of 10*g*; as a comparison, 7*g* is enough to cause blackouts and temporary blindness in fighter pilots. The reason for all this effort? It's simply to produce a squeaking sound with the tail feathers—the faster the bird, the louder the squeak. This may not sound like something worth risking life and limb for, but female hummingbirds are clearly impressed, favoring the most brazen masters of aerobatics.

Julia butterfly

Dryas iulia

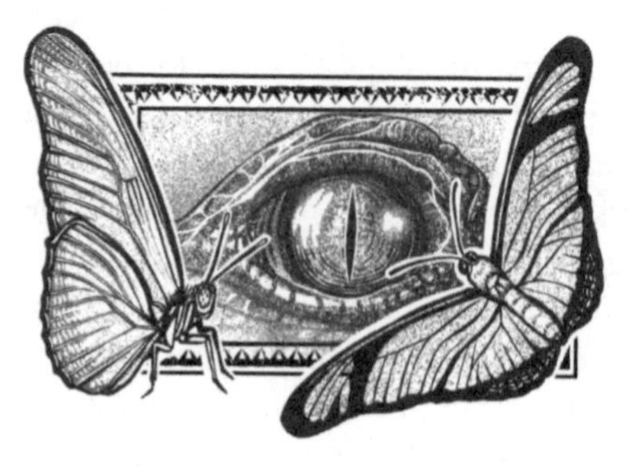

The phrase "to shed crocodile tears," which denotes a display of insincere emotion such as sympathy or sorrow, has been known since antiquity, dating back at least as far as Plutarch. But what would make a real crocodile weep?

While crocodilians are not usually known to be a blubbering mess, there is a creature that does cause them to cry in real life. That creature is the innocuous-looking Julia butterfly, *Dryas iulia*—and it does so to drink crocodile tears. The act, known as lachryphagy (translated as "tear-eating"), may seem like a very bizarre combination of meanness, grossness, and sheer nerve, but is merely a way of obtaining necessary minerals. The butterfly goes as far as poking the reptile's eyes with its proboscis to stimulate tear production. It's not just crocodilians (especially caimans) who fall victim to butterfly bullying—turtles are also a good source of nutritious tears.

Tear-drinking is part of a wider group of behaviors, termed "puddling"—or, in layman's terms, "drinking weird substances to obtain nutrients." Butterflies, as well as a few other insect species, will ingest the strangest liquids—from puddle water, rotting plant and animal matter, to dung, urine, and, quite literally, blood, sweat, and tears. In Julia butterflies, it is mainly males who engage in puddling—they require minerals to produce spermatophores (packets for storing sperm). These mineral-rich parcels serve as a nuptial gift when they are passed on to females during copulation. Lady Julias, meanwhile, stick to a vegetarian diet, and use nutrients from pollen to produce eggs. Both sexes feed on nectar as well.

The eggs are laid on passion vines—and the passion vine plants are not too thrilled at the prospect of being munched up by hungry Julia caterpillars. Consequently, various passion vine species have evolved defenses against herbivorous insects—rather intricate ones, too. The most basic protective mechanism is growing thick leaves that are hard to break down, as well as modified hooked leaf hairs that may pierce the body of a larva. A more sophisticated one is deception: some passion vines grow pretty convincing decoy butterfly eggs, to signal that they have already been occupied and that the laying female should move along. Some produce tendril-like growths that are seemingly the perfect laying spot—however, after the eggs are placed, the tendrils fall off, and the plant ditches the problematic freeloader. Passion vines contain chemical defenses as well, including trace amounts of cyanide that could act as insect deterrents. Moreover, most species will seek allies by producing a sugary substance, irresistible to ants. The ants, apart from having a sweet . . . well, mandible, will also attack and carry off butterfly larvae.

But butterflies can see through some of these tricks. Julia mothers, who "taste" with their feet because of the chemoreceptors located there, will scout out the laying sites very meticulously. At times, they may lay eggs next to, rather than on, the target plants, to protect the offspring from predatory ants, while still ensuring that food is only a short stroll away. Larvae will forage on tendrils rather than chewy leaves, and walk under or over the dangerous leaf hooks. As for the cyanide, it actually works to Julias' advantage, by making them unpalatable to their predators.

In fact, it is the cyanide in the passion vines that has become a recognition symbol for Julias and their kin; their vibrant orange wings serve as a warning signal: "I taste bad, avoid eating me." Julias are not the only species using orange as a heads-up—other unpalatable neotropical butterflies will do the same. This like-minded, honest signaling is called Müllerian mimicry, and butterflies forming the "orange complex" reinforce the message: stay away from the gingers!

Laysan albatross

Phoebastria immutabilis

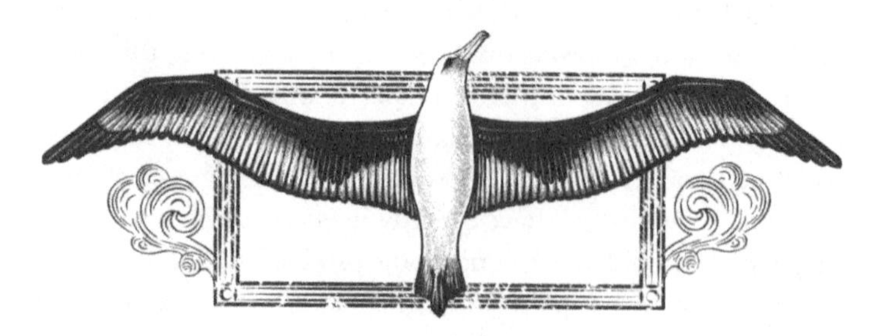

"Sleek, white-and-black-plumed, 2m-wingspan female seeking a serious relationship"—reads the dating profile of the Laysan albatross (*Phoebastria immutabilis*). "Likes dancing, seafood, and long travels. Based in Hawaii. Open to guys or gals."

When it comes to dating, albatrosses are all about commitment, an approach dictated by the challenges of their seabird lifestyle. Laysan albatrosses nest on remote oceanic islands—their breeding colonies are found almost exclusively in northwestern Hawaii—but, as surface-feeders, they need to fly across vast areas scanning oceanic waters for cephalopods. While indefinitely long voyages are fine for carefree singletons, raising a family requires a more predictable routine.

Albatrosses are ready for love when they are 5–7 years old. After a courtship that includes dancing (with elaborate poses and vocalizations), the happy couple is ready to breed. Once the egg is laid, both adults take turns sitting on the nest. Parenting is a two-bird job: eggs need to be incubated for about sixty-five days and chicks looked after for another three months—leaving them unattended for weeks would spell a death sentence.

As one parent incubates, the other goes on foraging trips covering several thousand kilometers. Laysan albatrosses incubating an egg in Hawaii have been found in Japan on their "days off." Meanwhile, as the off-duty parent flies to Hokkaido or another destination, the nest-bound parent is stuck, waiting. The longest recorded wait was fifty-eight days. Throughout their shifts birds don't move from the nest, don't feed, and only occasionally catch a few drops of rain to drink; they can lose over a fifth of their body weight during a stint. Such dedication requires a partnership for better or worse, richer or poorer and as long as both birds shall live. But what is an albatross to do if there aren't enough mates of the opposite sex to go around?

In 2008, Lindsay Young and colleagues reported that in a Laysan albatross colony in Oahu, Hawaii, where there is a shortage of bachelor males, 31 percent of breeding pairs consisted of two unrelated females sharing parental duties over a single egg. In most cases, the egg is fertilized by a paired male, indicating that even socially monogamous birds sometimes stray from the straight and narrow. If both mothers lay an egg in the same season, only one is incubated—the choice of whose it is appears to be random. While it may sound unfair, albatrosses are woefully unobservant and will happily incubate a coffee cup or a beer can placed in the nest in lieu of an egg. Incidentally, this inattention can be deadly when, instead of a delicious squid, they swallow a floating plastic bag.

Albatrosses generally mate for life, and, accordingly, successful female–female couples stick together for years, giving both females in the pair a turn to reproduce. Although the hatching rates are lower than in male–female pairs, the fledgling rates are comparable, indicating that two mothers can still rear a chick just fine. Despite a lower reproductive success than traditional couples, a female–female pairing is certainly better than not breeding at all. The all-female couples become rather intimate, with mutual preening and mate-guarding, indicating that they really are in it for the long haul.

In the case of Laysan albatrosses, the long haul can indeed be long. A female called Wisdom, banded on 10 December 1956 when her age was conservatively estimated at five, is the oldest confirmed wild bird in the world. Not only did she outlive Chandler Robbins, the researcher who banded her, but she has also probably taken on new mates after having been widowed. Wisdom shows no signs of stopping when it comes to babies, having hatched a chick (her thirtieth or fortieth) in February 2021, at the ripe age of seventy.

In a world full of traumatic inseminations, sneaky fuckers, and orgies inside sea cucumber bottoms (see pages 29, 32, and 126), it is good to remind oneself that romance and wholesome family values are not completely dead.

Marabou stork

Leptoptilos crumeniferus

The "Big Five"—buffalo, elephant, leopard, lion and rhino—are known as the most iconic African animals. However, few tourists would have heard of Africa's "Ugly Five"—the homeliest species on the continent: spotted hyena, warthog, lappet-faced vulture, wildebeest, and the marabou stork. While beauty might be in the eye of the beholder, that last animal certainly deserves its position in the pantheon of ugliness.

The marabou stork, *Leptoptilos crumeniferus*, is a huge bird, up to 150cm tall, and with one of the largest wingspans in the world—320cm, dwarfed only by the Andean condor and some albatrosses and pelicans. The sheer size is imposing, but what makes even more of an impression are the bird's looks. The marabou stork's hunched black back with a collar and front of white feathers gained it the nickname "the undertaker." The long, stick-thin legs are—theoretically—black; however, in practice they look white, as, in the effort to keep cool, the stork covers them in its own feces. The head, armed with an oversized bill, is bald. But not neatly, evenly bald like a vulture's; it is pockmarked with scabby, reddish patches, with occasional fuzzy feathers here and there—it looks as if the marabou had a serious accident involving a scalding wig. The neck gives "double chin" a new meaning—pink to magenta, naked and long, it has a dangly, wrinkled pouch (the gular sack) that can be inflated to assert dominance over other marabous.

Paradoxically, fashion lovers might associate marabous with their exquisitely beautiful down, used for fine sartorial trimmings, including the marabou mules worn by Marilyn Monroe in *The Seven*

Year Itch. Using marabou down for lingerie embellishments is rather appropriate: the delicate fluff comes from the bird's "undertail-coverts," or . . . butt feathers.

Even though the species is unsightly overall, one might wonder if perhaps the marabou's habits are more enticing. Perhaps it has a lovely personality and a great sense of humour? The short answer is: no. In the case of this stork, the disagreeable appearance has evolved specifically to match its revolting lifestyle.

The species, abundant throughout tropical Africa, is a scavenger—and not a picky one at that. Marabous associate with vultures and other carrion-eaters; it is said that the species will eat virtually any animal matter from termites to a dead elephant. Its featherless scalp is, supposedly, adapted to maintain better hygiene during feeding on larger carcasses; realistically, the powerful beak, head, and neck are constantly decorated with congealed blood and animal remains. The upside is, with so few feathers, it is more difficult for dinner leftovers to weigh the animal down.

During the breeding season, when the demand for protein is higher because of dependent chicks, marabous will hunt live prey. Their diet comprises fish, frogs, and rodents, but may also include crocodile eggs and hatchlings, as well as birds, in particular cormorants, pelicans, and flamingos. Marabous are also attracted to grass fires, and will march ahead of the fire front, catching anything trying to save itself from the flames.

Over the past decades, the wild carrion in the marabou diet has been increasingly replaced by man-made carrion. The birds frequent landfills, abattoirs and fisheries, and will eat anything they can find: feces, plastic, shoes, stockings, bits of metal . . . Their table manners are nonexistent: they swallow large (up to 600g) food chunks whole, let the digestive juices strip any nutrients, and regurgitate the rest. The most preposterous marabou meal must be a butcher's knife, which, covered in animal entrails, was snatched and swallowed, only to be found a few days later, spotless and clear of all blood and residue. Because of their bullet-proof stomachs, marabous play an important role as garbage collectors (not to be sneered at in the light of increased urbanisation in Africa)—although the long-term effects of their ingestion of plastics and metals are still unknown.

With all that we know about the marabous, these storks don't bring babies—they most likely eat them.

Moths

order Lepidoptera

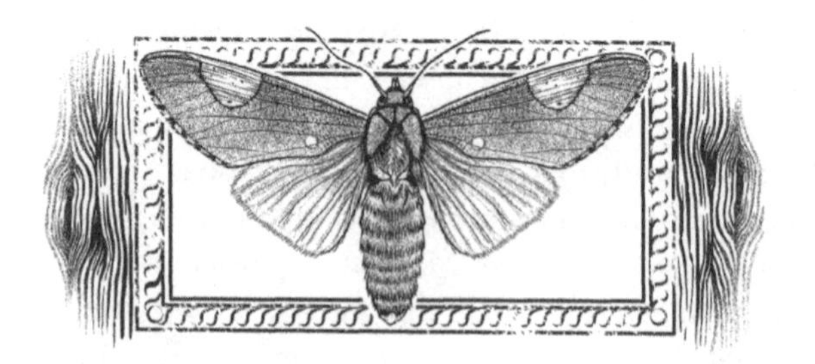

Warm summer nights are magically romantic and peaceful—that is, until you realize that they are actually full of blood-curdling screams, as loud as an approaching jet engine. Or perhaps you will not realize it, because—thankfully—these deafening shrieks are in the ultrasonic range (20–200kHz), too high for most humans to hear. The screams are emitted by bats, the banshees of the animal kingdom, when they echolocate to catch their flying dinners—moths.

To echolocate, a bat screams or clicks loudly, using its mouth or nose, and then listens out for the echo created when its calls bounce off objects; in this way, it builds an auditory "picture" of the world around it. When on a prowl, the bat emits long, exploratory calls, and upon detecting a tasty insect it will start a "feeding buzz"—an increased call rate—to home in on the prey and catch it.

But moths have a range of cunning defenses at their disposal. Alongside butterflies, moths belong to the order Lepidoptera; however, while butterflies have evolved from a common ancestor, moths do not form such a neat, monophyletic group—they are simply all the species of Lepidoptera not classed as butterflies. Yet, though they might be a taxonomic mess, they are good at escaping hungry bats.

Their most straightforward strategy is evasion. Moths may emerge in spring, when bats are less active, or become more diurnal, picking birds as the lesser of two evils. In flight, they can drop out of the bat's sonar beam by flying in loops, zig-zags or spirals. Or, on a larger evolutionary scale, they can arm themselves with a bat-detection system: ears (because ears are not part of a moth's basic package). These can be located on the mouth, thorax, or abdomen;

they consist of a membrane that vibrates in response to sound, and some auditory receptor cells that pick up the vibrations. With only 1–4 such cells per ear, moth ears are some of the simplest sensory organs in nature. But what they lack in sophistication, they make up for in functionality, allowing the insect to detect an approaching bat and either retreat or dodge it at the last moment.

Some moths go a step beyond evasion—they produce loud, distracting sounds to jam the bat signals. Tiger moths, such as *Bertholdia trigona*, interfere with oncoming bat sonar by clicking a part of the thorax called the tymbal, while male hawkmoths combine disruption and rudeness by grating their abdomens against special scraper cells on their genitals.

The jamming technique is also employed by moths without ears; the deaf *Yponomeuta* are known to click not just in response to bats, but perpetually. The earless moths' anti-bat protection is generally more passive; many species move more erratically to make themselves a less predictable target, while ghost moths, *Hepialus humuli*, fly close to vegetation, hiding in the clutter of background echoes.

Aside from ears and sound-producing structures, one more body part plays a role in dodging bats. Silk moths in the family Saturniidae excel at deception using wings. For instance, those of the earless Chinese tussar moth, *Antheraea pernyi*, are noise-canceling. Because they are covered in scales that absorb bat calls rather than reflecting them, the wings render their owner invisible to sonar. Other saturniids, however, have an even better solution, using long, thin tails embellishing their hind wings. These might look decorative, but they serve an important function: they're decoys, creating an echoic sensory illusion. As they trail behind the moth, they spin and divert the attention of bats to the non-essential appendages rather than the main body—a bit like acoustic tag rugby.

Bats, meanwhile, are also enhancing their predatory techniques: they may shriek at frequencies below the hearing range of moths, or call more quietly, essentially whispering as they approach a moth. This 65-million-year evolutionary arms race is set to run and run.

New Caledonian crow

Corvus moneduloides

In New Caledonia, an archipelago in the southwest Pacific Ocean, south of Vanuatu, lives a crow. It looks like most other crows—black, slender, with sleek, glossy feathers. Yet the New Caledonian crow (*Corvus moneduloides*) is not at all ordinary.

The corvids, members of the crow family, are considered some of the animal kingdom's brightest minds. In Japan, carrion crows (*C. corone*) have been observed to place nuts in front of cars to crack the hard shells. American crows (*C. brachyrhynchos*) remember people who have been mean to them—and not only do they scold the wrongdoers, they also tell other crows about them. Four-month-old ravens (*C. corax*) outcompete adult great apes when it comes to social and physical cognitive skills.

It is no wonder that corvids have been revered in a number of mythologies—Celts and Slavs believed crows had oracular powers, while ravens symbolized wisdom for Native Americans, reported news to the Norse god Odin, and were associated with Apollo, the Greek god of prophecy. Still, the New Caledonian crow, even though not firmly rooted in any particular lore, is exceptional.

Like other corvids, this crow is an omnivore—its diet includes fruit, nuts, seeds, eggs, insects, and snails (which it drops on to hard rocks to break open). The feature differentiating it from other crows is an atypical bill—a shorter, blunter and straighter one, with the lower mandible, very unusually, pointing up. This unique trait allows the New Caledonian crow to excel at using tools. Since New Caledonia does not have native woodpecker species, the crows occupy a similar ecological niche, probing wood for invertebrates. However, because they lack the powerful head and beak of the woodpeckers, they access this food supply via tools: sticks, barbed vines or leaves, which they use to pull out invertebrates. The straight bill allows a tighter grip on the stick, and enables the bird to get a better view of what it's doing, as its angle brings the tool into the range of binocular vision.

Using tools, while pretty neat, is not the most exceptional feature of the New Caledonian crows. If they cannot find an appropriate device, they will go ahead and make one themselves, by trimming sticks or ripping leaves to the correct length or shape. They use plants' natural features, such as thorns, to pull out food; they also clip forked twigs into a "tick" shape to make useful hooks. They are the only non-human animals to manufacture hooked tools in the wild.

Because of this ability, New Caledonian crows have attracted huge research interest. In captivity, these crafty birds will also make tools out of materials that don't occur naturally in the wild—for example, they are able to bend bits of wire into hooks, or trim cardboard into a suitable form. They are capable of producing complex tools when presented with short twigs and extension parts to join them together. They can also solve multi-step puzzles involving various objects to enable them to reach their food—the longest recorded one involved eight separate actions that had to be carried out in the correct sequence before the bird could release its reward.

Additionally, the New Caledonian crow understands water displacement. When presented with a water-filled tube containing an inaccessible, floating treat, it will drop stones and other objects into the tube until the water level is high enough for the bird to reach the food.

The question is: how do these cunning crows know how to make artisanal tools? They are not great at social learning: they don't seem to imitate each other or pick up skills from observing conspecifics. However, they appear to be able to replicate a tool by seeing the final product design—what's more, they have the ability to improve it. This makes sense; in the wild, young crows will spend time with their parents, borrowing their tools and using them regularly—and creating a mental image of a utensil that works well. They can then, in a bout of innovation, modify, and enhance such a tool, leading to the first sign of "cumulative cultural evolution"—the process of a society accumulating product improvements over time. Bird-brained? Not in the slightest.

Old World fruit bats

family Pteropodidae

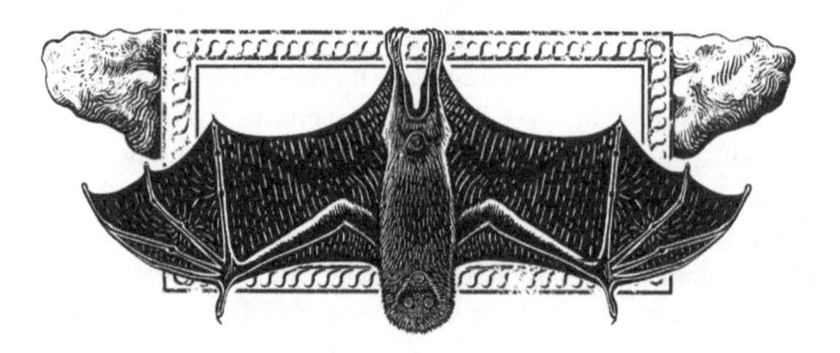

Megabat sounds like a great name for a new character in the *Batman* franchise. Still, with a hedonistic lifestyle and the ability to kill en masse, this bat might be better cast as supervillain than caped crusader.

Old World fruit bats are some 200 species of bat belonging to the family Pteropodidae, native to the tropics and subtropics of Africa, Eurasia, and Oceania. Though they are casually known as megabats, about a third of species are, in fact, not very mega at al—the smallest of the lot, the spotted-winged fruit bat, *Balionycteris maculata*, weighs a mere 13g (about 120 times less than some of the biggest species, collectively called flying foxes). Unlike the rather nightmarish-looking, echolocating, carnivorous microbats, most Old World fruit bats have pleasant, dog-like faces, are vegetarian, and find their way around with their keen eyesight, sense of smell and excellent spatial memory.

Fruit bats play an important role as seed dispersers, especially for plants with smaller seeds. Thanks to their size and a good set of teeth, larger bats can carry fruits at least as big as those swallowed by birds—and, once home, they are able to hang by one foot and use the other to manipulate their bounty. Even though most seeds don't spend long inside the bat (the rapid digestion of soft fruit prompts an exit within 10–70 minutes of entry), they can still be carried over distances of tens of kilometers. And when considering that social fruit bats can live in colonies of more than a million individuals (each pooping seeds all day, every day), it is no wonder that megabats are able to replant entire forests.

Meanwhile, fifteen or so species of Old World fruit bats are technically *nectar* bats, being specifically adapted to visiting flowers. Some, such as the Malaysian cave nectar bat, *Eonycteris spelaea*, travel up to 50km from their roosts for a floral meal. Long-snouted and long-tongued, they land on flowers to slurp up nectar and provide a pollinating service at the same time. Bat-pollinated blossoms are larger and more robust than insect- or bird-pollinated ones, and emit attractive scents at night to lure mammalian visitors.

What megabats also love to eat, apart from fruit and nectar, is each other. There have been reports of oral sex, both fellatio and cunnilingus, across different fruit bat species—before, during, and sometimes after copulation. The longer the foreplay or oral stimulation during sex (yes, they can bend like that), the longer the intercourse—and likely the higher the chance of fertilization. The Bonin flying fox, *Pteropus pselaphon*, displays fellatio between males—probably used as a means of avoiding conflict within a group.

Another interesting relationship is one with their pathogens. During flight, bat metabolism is very high, and body temperatures can jump to 41°C—which may cause DNA damage. While bats protect their own DNA with enhanced DNA repair pathways, resident viruses are less lucky. If they survive, they can stay—unlike other mammals, bats don't try to annihilate germs with a full-on inflammatory response, but are happy to let them linger at a low background level, merely limiting viral propagation. Consequently, bats harbor a range of pathogens without showing signs of disease—they are reservoirs of emerging viruses such as SARS, Middle Eastern Respiratory Syndrome and a range of other coronaviruses, as well as Nipah virus, Hendra virus, and, possibly, Ebola; they can also carry rabies.

Such tough conditions push the germs to evolve quickly—or perish. Problems start when these super pathogens infect species with less robust immune systems, like humans. Since large fruit bats are hunted for food across their entire geographic range, the risk of infection for people is high—and megabats, already threatened by habitat loss and food shortages, shed more pathogens if they are stressed or malnourished.

While fruit bats look sweet, have a sweet tooth and make sweet love, they are best left at a safe distance—because their revenge can also be sweet.

Orchid mantis

Hymenopus coronatus

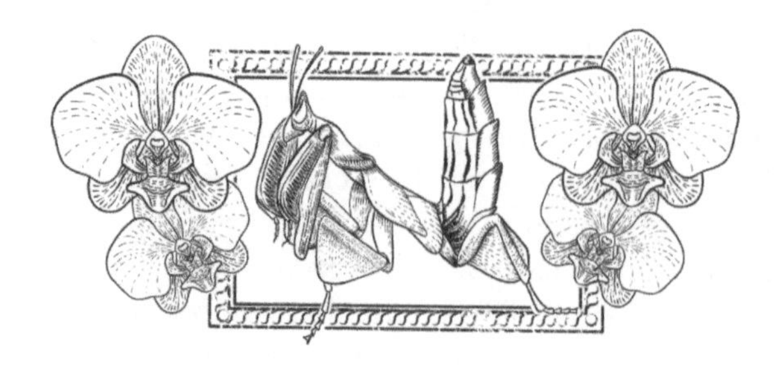

While touring Southeast Asia, the nineteenth-century Australian travel writer James Hingston was shown around a garden on the island of Java. His attention was drawn to a most astonishing flower: "a red orchid, that catches and feeds upon live flies." Hingston wrote in his account of the journey that the plant "seized upon a butterfly" and "enclosed it in its pretty but deadly leaves" as he watched. Yet what the writer saw was, in fact, not a murderous flower, but a strikingly accurate mimic.

Mimicry is one of evolution's niftier tricks. Protective mimicry, like that employed by the stick insect (see page 40), is great for blending in with the background to hide from dangers. The flip side, aggressive mimicry, is handy for making a predator invisible to prey. The orchid mantis, *Hymenopus coronatus*, uses both.

Functionally, the orchid mantis resembles other mantids: it is carnivorous, it has a creepy little triangular head, four walking legs and two grasping ones. But while most mantises are green or brown, this species puts the "pretty" in "pretty deadly"—it would not look out of place in a fancy corsage. The overall appearance is flamboyant: a delicate white or pink coloration, violet eyes, legs adorned with flattened lobes resembling petals, and an abdomen that the juvenile mantises bend upward to imitate an orchid even more closely. When sitting among other flowers, this mantis is indistinguishable from a blossom.

While many predatory species will try to imitate petal coloration to trap their prey, the orchid mantis is the only animal to mimic an entire flower. Even though it does not resemble any particular plant species, its disguise is a real head-turner: the mantis attracts more pollinators than a real orchid. Its imperfect mimicry, or looking like a generic—though highly appealing—flower, gives it a wider appeal that taps into the differing tastes of potential prey. To avoid getting in each other's way, the similarly formed male and female mantises show different predatory strategies. The little, 2cm males are ambush predators that blend in with their surroundings to pounce on their victims, whereas the large, 6–7cm-long females make themselves as conspicuous as possible and attract pollinators with their looks. Bees, flies, and butterflies will take a detour to pollinate the novel bloom—and then, of course, it is too late to move away from the deadly grasp of the beautiful beast.

The idea that orchid mantises use predatory mimicry was first proposed by Alfred Russell Wallace in 1877. Poor, gullible bees seem to have a particularly difficult relationship with orchids—while some flowers pretend to look like female insects to attract sex-obsessed male bees (see page 159), here is an insect looking like a flower to, once again, prey on their naivety. Like real orchids, an orchid mantis absorbs UV radiation, which makes it stand out against UV-reflecting foliage and is highly noticeable for bees and wasps with UV vision. What's more, the mantis not only looks convincing, but it smells highly attractive to bees: juveniles emit the same chemical concoction as that used in the communication of the oriental honeybee.

At the same time, floral outfits seem to work very well when fooling predators. Birds and lizards "interpret" the orchid mantis as a flower, due either to camouflage (blending in with other blossoms and thus becoming invisible) or to masquerade—imitating an inedible object. Or perhaps it's a bit of both.

The mantis is such a successful flower mimic because it is rarer than its model, the real orchid blooms. Because of this, both prey and predators are, statistically, much more likely to encounter flowers than a dangerous (though potentially tasty) insect, and the mantis can maintain an aura of mystique. Flower power is evidently best used in moderation.

Paradise tree snake

Chrysopelea paradisi

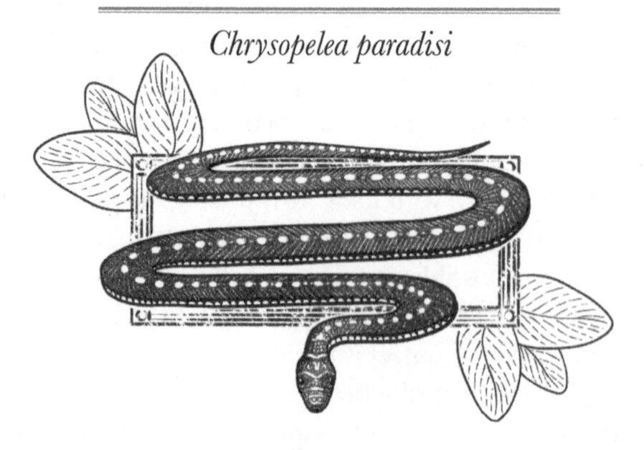

Is it a bird? Is it a plane? No, it's a flying snake! Well, technically more of a gliding snake, as the paradise tree snake, *Chrysopelea paradisi*, cannot ascend in the air. Still, it can cover a fair distance—over 30m in a horizontal line.

The paradise tree snake is one of five species of flying snakes; all are native to South and Southeast Asia. These aeronautic reptiles are small, measuring between 60 and 120cm; they weigh up to a few hundred grams. They are also splendidly colored, with yellowish-green scales and orange splotches that stand out against a black background. The arboreal snakes live and hunt in tree canopies, mainly feeding on lizards and bats—and use the hinged scales on their bellies for better grip when climbing. This might be hard work, but what is even more difficult is not falling down from a 75m tree once you get to the top.

For tree-dwelling creatures, gliding is an incredibly useful skill, as it helps prevent injuries from a fall or leap, and comes in handy during hunting or while escaping predators. However, being a limbless animal presents some aeronautical problems. The obvious one is the lack of wings, skin flaps, or other surface-enhancing appendages. The other one: there are no legs to kick off with during the start. Yet paradise tree snakes can land safely on the ground or vegetation after a leap from at least a 15m height—how do they do it?

The first method they employ is shapeshifting. To gain favorable aerodynamics, the reptile flattens itself and its cross-section changes from cylindrical to triangular. Upon departure, the snake tucks its belly in and spreads out its ribs, to make itself as ribbon-like as possible. The ribs move to the sides and forward, and stretch out the animal, starting from the head and finishing at its vent (bum); in the middle of the body, the snake is twice as wide as when at rest. This rib-stretching mechanism is the same as in cobras when they lift their hoods. The underside of the snake is almost concave, though some organs, such as the heart, protrude a bit from the overall smoothness—and any remaining, undigested food also spoils the sleek shape with small, irregular bumps on the surface of the belly. Because the ribs are busy pulling the snake into shape, it is likely that the reptile cannot breathe while gliding, limiting the duration of the flight.

The second tip for air travel is a good take-off. A suitable starting point is somewhere high up, ideally a branch. Paradise tree snakes cannot kick-start themselves into flying, so they either jump, dive, or fall. The latter is the easiest—all it takes is letting go from a hanging start. Dives involve undulating off a branch head-first, pointing downward. Yet the most common, and most efficient, method is the jump. The snake hangs on its tail from a perch in a "J" shape, pulls itself up and shoots into the air. This technique, while more time-consuming, increases the initial velocity and allows further travel than the other two starts.

The third secret of snake aeronautics is dynamic flight. These animals are not passive gliders—they use aerial undulation, slithering from side to side in mid-air, to become a morphing, flattened wing. The faster the undulations, the more stable the body; the smaller the snake, the faster the undulations—which is why petite snakes tend to be better at flying. They can also pilot themselves in mid-air, as their head remains relatively stable, and turn if need be.

If the thought of an undulating snake descending from a great height gives you nightmares, you can at least take small solace from the fact that they aren't very venomous to humans.

Periodical cicadas

Magicicada spp.

Loud, obnoxious, somewhat intimidating and impossible to reason with—cicadas are the worst wedding guests ever. When planning nuptials in northeastern United States, take them into account, because unless vibrating, red-eyed, orange-winged confetti is a special feature in your wedding plans, you might find yourself in more of a "marry in May, rue the day" situation.

Periodical cicadas (seven species in the genus *Magicicada*) start their lives inside twigs—as eggs. Within 6–8 weeks they hatch and migrate underground, where they feed on root sap until their teen years; an oddly precise number of years, too—thirteen or seventeen, depending on species. When nymphs are finally ready to resurface, they do so en masse, within a few days of each other. Emergence events are humongous: cicada densities reach 3.5 million per hectare, and the ground is literally crawling with insects. They emerge in spring, around May, prompted by the rising temperature of the soil; though it is unclear how they know so accurately which year it is. For the last few weeks of their lives, they transform into adult—to mate, lay eggs, and ruin special occasions with their incessant noise.

Bob Dylan's 1970 song about his degree ceremony in Princeton mentioned that the locusts were singing and they were singing for him. Dylan may win Nobel Prizes for his lyrics, but he's no entomologist. The "locusts" he mentions were in fact periodical cicadas; they didn't technically sing, but raucously vibrated their abdominal tymbals, and, rather than for him, they were doing it to attract mates. At over 95 decibels, the din is so loud it can cause permanent hearing damage in humans. Not only do they drown out speeches with their loud backsides, they will also urinate on the heads of guests—euphemistically called "cicada rain" or "honeydew." A small comfort is that they don't bite; like other true bugs, they only have sucking mouthparts.

Periodical cicadas appear in broods—populations emerging at the same time and in contiguous locations. In 1907, the American entomologist Charles Lester Marlatt gave the broods Roman numerals, I–XVII for the seventeen-year cicadas, and XVIII–XXX for the thirteen-year ones. Only fifteen broods are still alive today, and keeping track of them—as well as predicting their future prospects—requires a team effort. Consequently, periodical cicada research is a great example of how citizen science has changed over time.

Take Brood X, for instance—every single one of its emergences has been documented since 1715. In 1851, entomologist Gideon B. Smith published newspaper columns asking the public to contact him with reports of cicada appearances. In 1902, Marlatt and his colleagues at the US Department of Agriculture sent out 15,000 postcards soliciting emergence records. In 1987, universities ran telephone hotlines. In 2004, email was the go-to solution. Most recently, in 2021, people could use an app to log photo and video records along with the exact geographical location, allowing researchers to identify cicada species, ranges, and potential threats.

The periodical, mass emergence is in itself a survival strategy. The sheer quantity of adult insects means that no amount of predation can make a dent in the brood—predators are simply oversaturated. Meanwhile, prime-numbered breeding cycles make the cicadas an unreliable food source for any animal breeding in more regular patterns. It seems to work; only one known adversary managed to synchronize with the cicadas' strange periodicity: the fungus *Massospora cicadina*. But it did so with a vengeance. The fungus attacks only male cicadas, destroying their genitalia, while at the same time making the rest of the animal more sex-crazed than before. The male will not only try to mate with females (not to reproduce, since he is no longer able to do so, but to spread the fungal spores); he will also imitate the female wing-flicking sound to attract other males and pass on the infection to them. Periodical cicadas—noisy, horny, urinating everywhere, and ridden with STDs—are as insufferable as bachelor parties.

Regent honeyeater

Anthochaera phrygia

The key to linguistic proficiency is exposure: hearing, living, breathing a language. This holds true for the vocal culture of both humans and animals—for example, the songs of birds or whales. Animals pick up phrases and notes from each other, and unused song elements become forgotten and lost. Rare languages with very few speakers go extinct—but while the death of an obsolete human language is regrettable, in the case of birds it can indicate the disappearance of an entire species.

Such a sad trend is currently happening in front of our eyes—or, rather, ears—in the forests of southeastern Australia, the last refuge of regent honeyeaters, *Anthochaera phrygia.* These starling-sized, gold-speckled black songbirds used to be abundant until the mid-twentieth century, traveling in flocks of hundreds in search of nectar from flowering eucalyptus trees; their calls were regularly heard throughout cities and villages. Regent honeyeaters favor box-ironbark forests, abounding in trees that flower year-round, supporting a plethora of nectarivorous species. Unfortunately, since the 1940s, about 75 percent of honeyeater habitat has been cleared for housing and agricultural developments, and the once common species is now classified as critically endangered. The entire population has been reduced to 200–400 individuals, spread across a range of 300,000 square kilometers—an area roughly the size of Italy.

During breeding season, the males court the females with their vocal skills; their songs, like human language, are not innate, but learned. Young males don't learn vocalizations from their fathers, since adults don't sing while their young are still residing on their territories. Instead, the youngsters are reliant on copying other mature males later in life. However, with their nomadic lifestyles and extremely low densities, regent honeyeaters are struggling to find language tutors during their critical learning period, as it is not unusual for the nearest male to be hundreds of kilometers away.

What the young males resort to as an alternative is copying other species: they incorporate the songs of rosellas, friarbirds, currawongs ,or wattlebirds into their repertoires. This seems like a good move—after all, a number of birds mimic all sorts of sounds, the prime example being the lyrebird, *Menura novaehollandiae*, who can imitate anything from a kookaburra to a chainsaw (to the admiration of lyrebird females). Complex songs may serve as an honest signal of male quality—more ambitious solos have been associated with higher reproductive success and low parasite numbers in some species. The more sophisticated the song, the fitter the individual.

There is a catch, however. While we might think that speaking many languages is sexy, female regent honeyeaters don't agree. They are looking for mates who fulfill honeyeater cultural norms, and stick to traditional songs; males who "speak foreign" are much less likely to find a mate. Not only is this frustrating for the singletons, but it can be detrimental to honeyeater couples, as single males often disturb nest sites, hoping to get the females' attention even though they are clearly otherwise engaged.

A similar problem persists with regent honeyeaters bred in captivity. When released into the wild, birds using the "captive" jargon may have smaller chances of finding a mate among linguistic purists. Captive breeding programs have therefore started tutoring the young males using recordings of wild regent honeyeater songs. While the language boot camps have increased the birds' survival abilities in the wild (for reasons that are not yet fully understood), the jury is still out on whether they have helped with attracting the opposite sex.

While it is apparent that there is a correlation between song type and reproductive fitness, it does not necessarily imply causation. Still, regent honeyeaters are so sparsely distributed that finding a tutor male is like looking for a needle in a haystack. Analyzing song characteristics can be a good predictor of bird density and therefore the state of the population—which, at the moment, is sadly woeful. To ensure the survival of the regent honeyeater species, we must not meet their plea with silence.

Sociable weaver

Philetairus socius

New developments keep popping up across the Kalahari Desert in southern Africa—spacious, comfortable lodgings, with countless apartments up for grabs. Get in while you can!

The developers—and builders—behind these investments are sociable weavers, *Philetairus socius*; brown, sparrow-sized birds. Although they look inconspicuous, they produce the most spectacular structures of all bird engineers: huge, complex nests, with several hundred individual chambers. These avian estates are usually placed on trees (although man-made structures, such as electricity poles, are also desirable), can measure up to 6m across and weigh a ton. The nests, made of twigs and grasses, look a bit like haystacks—or oversized shaggy bears perched on a tree. They are covered with a thatched "roof," and the small entrances, accessible from below, form a honeycomb of openings, each around 5cm wide, leading to private quarters. These permanent constructions last over a hundred years; they are inhabited by multiple generations of sociable weavers, and the individual flats, lined with softer materials, are where a weaver couple will breed and rest at night.

Building and maintaining such a grand property is no easy feat, and requires collaboration within the colony. Yet some birds tend to focus their efforts on looking after their individual apartments, rather than helping with the communal areas, such as the thatch. Shirking collective responsibilities does not go unnoticed: the weavers who put more time and energy into patching the shared roof will become very aggressive toward the selfish freeloaders, and chase them away. The bullying works—after having been chased, the lazy birds make more effort to contribute toward the upkeep of public spaces. This system is called the "pay-to-stay" model, as individuals become more collaborative out of fear of being evicted permanently.

These architectural experts fall into a category of animals called "ecosystem engineers"—species that significantly influence the availability of resources, such as food or shelter, in their environment. The communal nests are more than apartment complexes—they are whole districts, attracting not just weavers, but a host of other creatures. In fact, trees with weaver colonies draw thirty-six times as many species as non-colony trees. The soils below the nests are very nutrient-rich (no surprise, with so many occupants providing daily fertilizer), which prompts richer vegetation growth and a reliable food supply for ungulates in an unpredictable climate. Large herbivores will seek shade under the nest-canopies; carnivores, such as cheetahs or leopards, exploit them as lookout platforms, and large raptors, such as vultures, will build their own nest on top of the thatches. Because of their prominence within the landscape, trees with weaver colonies are often used as landmarks for scent-marking as part of territorial behavior.

Yet what is particularly important for any egg-laying creatures is the stable temperature the nests offer—a rare commodity in the harsh desert environment. The Kalahari can be as hot as 45°C; however, the well-insulated nests protect broods from overheating. Similarly, in winter, when temperatures fall below zero, the chambers offer a refuge for birds that otherwise risk hypothermia. The deepest lodgings provide the best insulation, being up to 24 percent cooler than the ambient temperatures in summer and three times warmer in winter, and are thus occupied by the dominant weavers in the colony. Yet the nests attract lodgers from a number of species, including lovebirds, owls, finches, and falcons.

Some of these occupants are problematic. African pygmy falcons, *Polihierax semitorquatus*, never bother to build their own nests—they just barge into a weaver colony and settle there. They may even have multiple homes in multiple colonies. Adding insult to injury, these ungrateful tenants will at times predate upon their hospitable hosts. Weavers raise alarm calls whenever a falcon appears—and other lodgers threatened by the raptors, such as the Kalahari tree skinks (*Trachylepis spilogaster*), will eavesdrop on these signals and react accordingly. Weaver nests really are the perfect homes: well insulated, rent-free—and they even come with their own burglar alarms.

Vampire finch

Geospiza septentrionalis

If you find yourself marooned on an inhospitable island, follow the example of Robinson Crusoe: make the most of what is around you and be adaptable. The forebears (forebirds?) of Darwin's finches have done just that.

Some 1.5 million years ago, ancestral finches from South America found their way to the Galápagos Islands, almost 1,000km west of continental Ecuador. The castaways gave rise to eighteen species of Darwin's finch (technically not true finches, but tanagers classified in the subfamily Geospizinae), which currently inhabit the archipelago. Present-day Darwin's finches are all small and drab in color, but show an astounding diversity of beak sizes and shapes. These beaks, ranging from the huge, broad, and blunt bill of the large ground finch to the petite, narrow bill of the warbler finch, are adapted for distinct diets. While broad-billed birds thrive on islands with nuts, finches with longer and more slender bills take advantage of sites with cactuses, and those with the smallest beaks specialise on insects. Each type of finch forages for a slightly different kind of food—biologists would say that each species occupies a different ecological niche—and consequently avoids getting in the way of the others, so much so that over time the populations with distinct beaks became separate species. Darwin's finches are therefore a model example of adaptive radiation, a phenomenon where species diverge from a common ancestor as they adapt to different ecological opportunities—and a diagram of the birds and their varied beaks is a staple fixture in evolution textbooks.

But while ground finches feed on seeds, and tree finches on insects, one type of finch, inhabiting the two most remote islands,

Wolf and Darwin, takes a more gruesome path. Originally thought to be a subspecies of the insectivorous sharp-beaked ground finch, this bird has recently been elevated to its own species: the vampire finch (*Geospiza septentrionalis*). During the wet season, when food is abundant, vampire finches will happily forage on insects, seeds or nectar. However, in the dry season, the lack of food and drinking water drives the birds to switch to more grisly sustenance: blood. This iron-rich snack forms such a significant part of their diet that vampire finches acquired a specialized gut flora, containing bacteria usually found in the intestines of carnivorous birds and reptiles.

For its sanguineous meal, the audacious little bird targets larger species, particularly red-footed and Nazca boobies (see page 162). The dainty vampires perch on the boobies' rumps and draw blood by pecking with scalpel-like beaks at the base of the wing feathers. The flowing blood attracts more finches, and a queue forms around the victim—everyone wants a sip. Adult boobies are not pleased about it, but seem to mind the vampire finches less than the flies their wounds attract. Young, fluff-covered boobies have a much rougher time. The finches peck relentlessly at the softest body part, the cloaca, sometimes causing the baby boobies to flee their nests—with fatal consequences.

In all likelihood, this relationship started off differently—as a mutualistic arrangement, with boobies letting finches peck flies and lice off their backs. Both sides were happy: the finches got food and the booby got rid of parasites, much like the cleaner wrasses and their clientele (see page 92). But trying conditions create tough birds—and the relationship soured once the finches realized that instead of invertebrates, an unreliable food source, they could have a liquid meal on tap. The birds gave into temptation and, just like that, mutualism turned into parasitism.

To make matters worse, vampire finches also feed on booby eggs—a remarkable undertaking when one realizes that the egg weighs more than twice as much as the 20g diner. The little bird uses its extra-sharp beak to pierce the shell, or it can kick the well-packaged lunch into a rock or over a cliff to open it. The initial effort pays off, since the egg provides excellent sustenance.

Thankfully, Robinson Crusoe did not have to resort to becoming a vampire—but perhaps, given a few hundred thousand years, his descendants would have?

White butterfly parasite wasp

Cotesia glomerata

Are you a screenwriter looking for a plot for a B-movie, perhaps sci-fi or horror? Look no further, nature comes to the rescue.

The story opens with the victim—a butterfly. Specifically, the cabbage white butterfly, *Pieris rapae*, found commonly throughout Europe, Asia, and North Africa. More precisely—its larval form, a caterpillar, enjoying a sunny day of munching Brussels sprouts in a garden. Enter an innocent-looking black wasp, only 3–7mm long, akin to a flying ant. The wasp looks inoffensive, but the music foreshadows a threat. And indeed—the wasp lands on the caterpillar, pierces it with its sharp ovipositor and proceeds to lay a few dozen eggs inside the living, eating, cabbage white. It's the antagonist: a female white butterfly parasite wasp, *Cotesia glomerata*—a parasitoid. Unlike parasites, who don't usually kill their hosts, parasitoids don't care if their chosen caterpillar lives or dies (spoiler: it dies). The white butterfly parasite wasp is a koinobiont; it does not slay the caterpillar immediately, but lets it live, grow, and even undergo metamorphosis—a sensible move, since a bigger host means more food for young wasps. Two to three weeks later, wasp larvae emerge, killing the caterpillar and building cocoons on its remains, like true body-snatchers. An unparasitised caterpillar turns into a beautiful, white butterfly—a parasitised one disintegrates into a bag of wasp larvae.

Parasitoidy is very common among wasps (see Emerald cockroach wasp, page 180), as tens, if not hundreds, of thousands of species employ this reproductive strategy. And, to be fair, the motive is also not uncommon in sci-fi movies (see Ridley Scott's *Alien*). But there is more to our plot: once the eggs are laid in the unsuspecting caterpillar, in comes the avenger. It's another wasp, *Lysibia nana*, who—you guessed it—lays eggs inside the larvae of the white butterfly parasite wasp. Unlike the original parasitoid wasp, this hyperparasitoid, a parasitoid of parasitoids, is what's known as an idiobiont: in other words, it paralyzes the host straightaway via venom injection, preventing any further development. It then lays a single egg in an existing wasp larva, creating a rather morbid larval Russian doll. When the egg hatches, the hyperparasitoid wasp larva pierces the skin of its host and sucks out its insides, eventually eating all of it and moving into its cocoon to pupate. Because it takes up all the available space, the adult forms of the hyperparasitoid *L. nana* and parasitoid *C. glomerata* are surprisingly similar in size.

Now for the final twist—who orchestrated all of this? Who is the mastermind behind the psychopathic killings? It's a good one, because nobody saw it coming. The evil genius is . . . the plant. Yes, when chewed on by the cabbage white butterfly caterpillars, that Brussels sprout bush (or, frankly, any cabbage relative could be cast in this role) emits a chemical cry for help. Attacked plants release substances called herbivore-induced plant volatiles, botanical equivalents of a bat-signal. Parasitoid wasps sense these scents and fly over to parasitise the hungry caterpillars. But once parasitised, the caterpillars change; what changes, too, is the composition of their oral secretions. As a consequence, the volatile emissions of the plant chewed up by a parasitised herbivore smell different from those emitted by an unparasitised animal. The difference in smell is enough to inform the hyperparasitoid of the presence of freshly laid wasp eggs. In the end, two winners arise: the plant and the newly emerged hyperparasitoid wasp.

As the camera pans over the beautiful garden landscape, the narrator's voice (perhaps Morgan Freeman's?) recites an excerpt from Jonathan Swift's *On Poetry: A Rhapsody*:

> So, nat'ralists observe, a flea
> Hath smaller fleas that on him prey;
> And these have smaller fleas to bite 'em;
> And so proceed *ad infinitum.*

Credit roll. The end.

White-rumped vulture

Gyps bengalensis

In the 1980s, Madonna was just becoming popular in the US, Margaret Thatcher was prime minister of the UK, and white-rumped vultures, *Gyps bengalensis*, soared over India in unbelievably high numbers. The population of these bulky, bald-headed birds was estimated at over 10 million; in fact, they were considered the commonest large raptor in the world. Delhi averaged three white-rumped vulture nests per square kilometer, and national parks four times as many. As scavengers feeding on large ungulates, the birds found India, a country full of cattle, a paradise on Earth.

Vulture presence benefits people—flocks of birds that eat half a kilo of meat per day each clear away cow remains in no time, preventing the spread of disease and water-source tainting. The vultures' digestive system is nothing short of invincible—by eating infected carcasses, they help control livestock illnesses such as anthrax, brucellosis, or tuberculosis. Ten million birds equals the disposal of 5 million kilograms of meat per day—a sizeable sanitation effort. Additionally, vultures play a more spiritual role in the funerary traditions of Zoroastrians and Tibetans, who leave their dead to be carried away by the birds as part of a "sky burial."

Yes, the 1980s were a great time to be alive if you were a vulture in Asia.

Fast-forward to the year 2003. Madonna has just released her ninth studio album, the UK under Tony Blair began its invasion of Iraq, and the population of the white-rumped vultures fell by a shattering 99.7 percent. The causes of such a catastrophic decline were a mystery—especially since other vultures were also hit hard; for example, the slender-billed vulture (*G. tenuirostris*) populations fell by 97.4 percent. Alarmed scientists explored many options—disease, lack of prey, persecution, and environmental pollutants such as pesticides among them—but nothing seemed to fit. Eventually it was discovered that the culprit behind the unprecedented mortality was diclofenac, a non-steroid anti-inflammatory drug used on cattle. While not harmful to mammals, it causes kidney failure and death in vultures—and because multiple birds feed on the same cadaver, only a small fraction of diclofenac-treated cattle carcasses is enough to kill off a disproportionately large number of scavengers.

The loss of vultures opened up food resources to other scavenging species, such as feral dogs or rats. Unfortunately, while vultures are a dead-end for pathogens, the mammalian carrion-eaters become their reservoirs, additionally carrying diseases such as rabies. What's more, increased numbers of dogs in cities attract leopards, leading to human–wildlife conflict. Overall, the lack of vultures is exacerbating public health problems across the subcontinent.

In 2006, diclofenac was banned for use in cattle in India, Pakistan, and Nepal; Bangladesh followed in 2010. Three years after the ban, the use of the drug had dropped, but not completely—and vultures still died because of the drug-induced renal failure (though deaths occurred at a lower rate than before). A new and vulture-safe alternative, meloxicam, has since been developed; however, the uptake is still not prevalent. Moreover, because diclofenac is prescribed to humans, it sometimes trickles down to be used in veterinary contexts, though the human equivalent is now sold only in small vials to make buying large quantities unprofitable.

Meanwhile, vultures are slow-breeding birds, laying a single egg at a time, and while currently the decline in numbers has ceased, there is no measurable recovery yet. White-rumped vultures are still classified as critically endangered, with an estimated 3,500–15,000 individuals left.

Worryingly, diclofenac has been permitted in veterinary use in Spain and Italy, on the grounds that a more meticulous carcass disposal methodology would prevent any contamination in one of the four vulture species found across southern Europe. Alas, the first report of a diclofenac-induced death—of the cinerous vulture, *Aegypius monachus*—has been reported in 2021. It seems the vultures still have a long way to go before they return to their glory days.

Zebra finch

Taeniopygia guttata

While human children will sometimes eavesdrop on their parents, especially when topics such as presents or birthday surprises are concerned, some elegant little gray-and-orange birds are one step ahead: they eavesdrop on their folks while still inside their eggs. These birds are the zebra finches, *Taeniopygia guttata*, small, seed-eating songbirds originating in Australia and Indonesia. They have become known around the world as popular pets, and free-living populations have been introduced to Portugal and Puerto Rico.

Zebra finches breed in large colonies, of up to 230 individuals. In such numbers, it is very tempting to saddle someone else with the burden of childcare (a phenomenon known as conspecific brood parasitism)—so tempting that between a third and a half of females will try to occasionally lay in someone else's nest. After all, chances are the deceived parents won't even notice the extra egg; every zebra finch egg looks very much the same, white or pale bluish-gray, with no markings. Yet it turns out that birds can recognize their own eggs using their sense of smell, and may abandon the nest if they spot an intruder egg early in their laying sequence.

If there is nothing suspicious about the clutch and the parents proceed with incubation, they will aim to provide the chicks with the best possible chances of survival. In a place which is particularly prone to fluctuating temperatures, growing up small may provide an advantage during hot weather—smaller bodies are easier to cool, and require less water. But, if it is likely that their birthday falls during a heat wave, is it possible to influence the size of the young once they are already developing inside an egg? Apparently, yes.

When temperatures are high (usually above 35°C), adult zebra finches will emit "heat calls"—fast, rhythmic songs, typically combined with panting as they try to cool themselves down. It turns out that embryos inside eggs can eavesdrop on these vocalizations—and modify their development accordingly, especially as their hatching date approaches. If youngsters hear the heat call, they grow up smaller than those with no such vocalizations in the background. This alteration appears to pay off: females who were little while growing up under hot conditions (or were bigger when reared in cold conditions) produced more offspring during their first breeding season than those not adapted to ambient temperatures. Additionally, birds who heard the call as embryos were more likely to vocalize themselves when experiencing high temperatures in the nest.

Heat calls are not the only vocalizations useful in zebra finch parenting. Since birds of this species mate for life, and are rather diligent at splitting parental duties (building nests, incubating, feeding the young), they need a means of negotiating who does what when. Finch moms and dads do so through song—a specific duet, performed during swap-overs between incubation shifts. When relieving each other from nest duty, they use the call—softer and more private than their usual chirps—to coordinate future workloads. In an experiment where males were delayed in returning to the nest, the structure of the duet performed upon their return changed: the song became shorter and more intense (possibly along the lines of "It's! Your! Turn! Now! Where! Have! You! Been?!?"). Following such a rapid duet, females would spend less time incubating during their subsequent shift. It's only fair!

References

INTRODUCTION

Cox, P., 'The *Physiologus*: A *Poiesis* of Nature', *Church History*, 52(4), 1983, pp. 433–43.

Giribet, G. and Edgecombe, G.D., *The Invertebrate Tree of Life*, Princeton University Press, 2020.

Harrison, P., 'The Bible and the emergence of modern science', *Science and Christian Belief*, 18(2), 2006, p. 115.

McCarthy, D.P., Donald, P.F., Scharlemann, J.P., Buchanan, G.M., Balmford, A., Green, J.M., Bennun, L.A., Burgess, N.D., Fishpool, L.D., Garnett, S.T. and Leonard, D.L., 'Financial costs of meeting global biodiversity conservation targets: current spending and unmet needs', *Science*, 338(6109), 2012, pp. 946–9.

EARTH

Antechinuses

Mason, E.D., Firn, J., Hines, H.B. and Baker, A.M., 'Breeding biology and growth in a new, threatened carnivorous marsupial', *Mammal Research*, 62(2), 2017, pp. 179–87.

Smith, G.C., Means, K. and Churchill, S., 'Aspects of the ecology of the Atherton antechinus (*Antechinus godmani*) living in sympatry with the rusty antechinus (*A. adustus*) in the Wet Tropics, Queensland – a trapping and radio-tracking study', *Australian Mammalogy*, 40(1), 2018, pp. 16–25.

Aye-aye

Erickson, C.J., 'Tap-scanning and extractive foraging in aye-ayes, *Daubentonia madagascariensis*', *Folia Primatologica*, 62(1–3), 1994, pp. 125–35.

Gochman, S.R., Brown, M.B. and Dominy, N.J., 'Alcohol discrimination and preferences in two species of nectar-feeding primate', *Royal Society Open Science*, 3(7), 2016, p. 160217.

Simons, E.L. and Meyers, D.M., 'Folklore and beliefs about the aye aye (*Daubentonia madagascariensis*)', *Lemur News*, 6, 2001, pp. 11–16.

Banana slugs

Chan, B., Balmforth, N.J. and Hosoi, A.E., 'Building a better snail: lubrication and adhesive locomotion', *Physics of Fluids*, 17(11), 2005, p. 113101.

Leonard, J.L., Pearse, J.S. and Harper, A.B., 'Comparative reproductive biology of *Ariolimax californicus* and *A. dolichophallus* (Gastropoda; Stylommiatophora)', *Invertebrate Reproduction and Development*, 41(1–3), 2002, pp. 83–93.

Mead, A.R., 'Revision of the giant West Coast land slugs of the genus *Ariolimax* Moerch (Pulmonata: Arionidae)', *The American Midland Naturalist*, 30(3), 1943, pp. 675–717.

Bat-eared fox

Clark, H.O., '*Otocyon megalotis*', *Mammalian Species*, 2005(766), 2005, pp. 1–5.

Stenkewitz, U. and Kamler, J.F., 'Birds feeding in association with bat-eared foxes on Benfontein Game Farm, South Africa', *Ostrich-Journal of African Ornithology*, 79(2), 2008, pp. 235–7.

Wright, H.W.Y., 'Paternal den attendance is the best predictor of offspring survival in the socially monogamous bat-eared fox', *Animal*, 71(3), 2006, pp. 503–10.

Brown rat

Bartal, I.B.A., Decety, J. and Mason, P., 'Empathy and pro-social behavior in rats', *Science*, 334(6061), 2011, pp. 1427–30.

Crawford, L.E., Knouse, L.E., Kent, M., Vavra, D., Harding, O., LeServe,

D. and Lambert, K.G., 'Enriched environment exposure accelerates rodent driving skills', *Behavioural Brain Research*, 378, 2020, p. 112309.
Quinn, L., Schuster, L.P., Aguilar-Rivera, M., Arnold, J., Ball, D., Gygi, E. and Chiba, A.A., 'When rats rescue robots', *Animal Behavior and Cognition*, 5(4), 2018, pp. 368–79.
Sato, N., Tan, L., Tate, K. and Okada, M., 'Rats demonstrate helping behavior toward a soaked conspecific', *Animal cognition*, 18(5), 2015, pp. 1039–47.

Caecilians

Kupfer, A., Müller, H., Antoniazzi, M.M., Jared, C., Greven, H., Nussbaum, R.A. and Wilkinson, M., 'Parental investment by skin feeding in a caecilian amphibian', *Nature*, 440(7086), 2006, pp. 926–9.
Measey, G.J. and Gaborieau, O., 'Termitivore or detritivore? A quantitative investigation into the diet of the East African caecilian *Boulengerula taitanus* (Amphibia: Gymnophiona: Caeciliidae)', *Animal Biology*, 54(1), 2004, pp. 45–6.
Wilkinson, M., Kupfer, A., Marques-Porto, R., Jeffkins, H., Antoniazzi, M.M. and Jared, C., 'One hundred million years of skin feeding? Extended parental care in a Neotropical caecilian (Amphibia: Gymnophiona)', *Biology Letters*, 4(4), 2008, pp. 358–61.
Wilkinson, M., Sherratt, E., Starace, F. and Gower, D.J., 'A new species of skin-feeding caecilian and the first report of reproductive mode in *Microcaecilia* (Amphibia: Gymnophiona: Siphonopidae)', *PLoS One*, 8(3), 2013, p. e57756.

Coconut crab

Drew, M.M., Harzsch, S., Stensmyr, M., Erland, S. and Hansson, B.S., 'A review of the biology and ecology of the robber crab, *Birgus latro* (Linnaeus, 1767) (Anomura: Coenobitidae)', *Zoologischer Anzeiger-A Journal of Comparative Zoology*, 249(1), 2010, pp. 45–67.
Laidre, M.E., 'Coconut crabs', *Current Biology*, 28(2), 2018, pp. R58–R60.
Oka, S.I., Tomita, T. and Miyamoto, K., 'A mighty claw: pinching force of the coconut crab, the largest terrestrial crustacean', *PloS One*, 11(11), 2016, p. e0166108.

Common bed bug

Polanco, A.M., Miller, D.M. and Brewster, C.C., 'Survivorship during starvation for *Cimex lectularius* L.', *Insects*, 2(2), 2011, pp. 232–42.
Reinhardt, K. and Siva-Jothy, M.T., 'Biology of the bed bugs (Cimicidae)', *Annual Review of Entomology*, 52, 2007, pp. 351–74.
Saenz, V.L., Booth, W., Schal, C. and Vargo, E.L., 'Genetic analysis of bed bug populations reveals small propagule size within individual infestations but high genetic diversity across infestations from the eastern United States', *Journal of Medical Entomology*, 49(4), 2012, pp. 865–75.
Smith, W., *A Dictionary of Greek and Roman Antiquities*, Harper & brothers, 1857.
Szalanski, A.L., Austin, J.W., McKern, J.A., McCoy, T., Steelman, C.D. and Miller, D.M., 'Time course analysis of bed bug, *Cimex lectularius* L. (Hemiptera: Cimicidae) blood meals with the use of polymerase chain reaction', *Journal of Agricultural and Urban Entomology*, 23(4), 2006, pp. 237–41.

Common sexton beetle

Andrews, C.P. and Smiseth, P.T., 'Differentiating among alternative models for the resolution of parent–offspring conflict', *Behavioral Ecology*, 24(5), 2013, pp. 1185–91.
Eggert, A.K., Reinking, M. and Müller, J.K., 'Parental care improves offspring survival and growth in burying beetles', *Animal Behaviour*, 55(1), 1998, pp. 97–107.
Scott, M.P., 'The ecology and behavior of burying beetles', *Annual review of entomology*, 43(1), 1998, pp. 595–618.
von Hoermann, C., Steiger, S., Müller, J.K. and Ayasse, M., 'Too fresh

is unattractive! The attraction of newly emerged *Nicrophorus vespilloides* females to odour bouquets of large cadavers at various stages of decomposition', *PLoS One*, 8(3), 2013, p. e58524.

Common side-blotched lizard

Corl, A., Davis, A.R., Kuchta, S.R. and Sinervo, B., 'Selective loss of polymorphic mating types is associated with rapid phenotypic evolution during morphic speciation. *Proceedings of the National Academy of Sciences*, 107(9), 2010, pp.4254–9.

Sinervo, B. and Clobert, J., 'Morphs, dispersal behavior, genetic similarity, and the evolution of cooperation', *Science*, 300(5627), 2003, pp. 1949–51.

Sinervo, B. and Lively, C.M., 'The rock–paper–scissors game and the evolution of alternative male strategies', *Nature*, 380(6571), 1996, pp. 240–43.

European rabbit

Hirakawa, H., 'Coprophagy in leporids and other mammalian herbivores', *Mammal Review*, 31(1), 2001, pp. 61–80.

Jernelöv, A., 'Rabbits in Australia' in *The Long-Term Fate of Invasive Species*, Springer, Cham, 2017, pp. 73–89.

Macdonald, D.W., *The Encyclopedia of Mammals*, Oxford University Press, 2006.

Face mites

Jarmuda, S., O'Reilly, N., Żaba, R., Jakubowicz, O., Szkaradkiewicz, A. and Kavanagh, K., 'Potential role of Demodex mites and bacteria in the induction of rosacea', *Journal of Medical Microbiology*, 61(11), 2012, pp. 1504–10.

Lacey, N., Raghallaigh, S.N. and Powell, F.C., 'Demodex mites – commensals, parasites or mutualistic organisms?', *Dermatology*, 222(2), 2011, p. 128.

Giant panda

Heiderer, M., Westenberg, C., Li, D., Zhang, H., Preininger, D. and Dungl, E., 'Giant panda twin rearing without assistance requires more interactions and less rest of the mother – a case study at Vienna Zoo', *PLoS One*, 13(11), 2018, p. e0207433.

Wei, R., Zhang, G., Yin, F., Zhang, H. and Liu, D., 'Enhancing captive breeding in giant pandas (*Ailuropoda melanoleuca*): maintaining lactation when cubs are rejected, and understanding variation in milk collection and associated factors', *Zoo Biology* (published in affiliation with the American Zoo and Aquarium Association), 28(4), 2009, pp. 331–42.

Zhang, G., Swaisgood, R.R. and Zhang, H., 'Evaluation of behavioral factors influencing reproductive success and failure in captive giant pandas', *Zoo Biology* (published in affiliation with the American Zoo and Aquarium Association), 23(1), 2004, pp. 15–31.

Zhu, L., Wu, Q., Dai, J., Zhang, S. and Wei, F., 'Evidence of cellulose metabolism by the giant panda gut microbiome', *Proceedings of the National Academy of Sciences*, 108(43), 2011, pp. 17714–19.

Giant prickly stick insect

Bian, X., Elgar, M.A. and Peters, R.A., 'The swaying behavior of *Extatosoma tiaratum*: motion camouflage in a stick insect?', *Behavioral Ecology*, 27(1), 2016, pp. 83–92.

Brock, P.D. and Hasenpusch, J.W., *The Complete Field Guide to Stick and Leaf Insects of Australia*, CSIRO Publishing, 2009.

Zeng, Y., Chang, S.W., Williams, J.Y., Nguyen, L.Y.N., Tang, J., Naing, G., Kazi, C. and Dudley, R., 'Canopy parkour: movement ecology of post-hatch dispersal in a gliding nymphal stick insect, *Extatosoma tiaratum*', *Journal of Experimental Biology*, 223(19), 2020, p. jeb226266.

Iwasaki's snail-eater

Danaisawadi, P., Asami, T., Ota, H., Sutcharit, C. and Panha, S., 'A snail-eating snake recognizes prey handedness', *Scientific Reports*, 6(1), 2016, pp. 1–8.

Hoso, M., Asami, T. and Hori, M., 'Right-handed snakes: convergent evolution of asymmetry for functional specialization', *Biology Letters*, 3(2), 2007, pp. 169–73.

Sheehy, C.M., 'Phylogenetic relationships and feeding behavior of Neotropical snail-eating snakes (Dipsadinae, Dipsadini)', PhD Dissertation, University of Texas, 2013.

Jumping spider

Chen, Z., Corlett, R.T., Jiao, X., Liu, S.J., Charles-Dominique, T., Zhang, S., Li, H., Lai, R., Long, C. and Quan, R.C., 'Prolonged milk provisioning in a jumping spider', *Science*, 362(6418), pp. 1052–5.

Huang, J.N., Cheng, R.C., Li, D. and Tso, I.M., 'Salticid predation as one potential driving force of ant mimicry in jumping spiders', *Proceedings of the Royal Society B: Biological Sciences*, 278(1710), 2011, pp. 1356–64.

Richman, D.B. and Jackson, R.R., 'A review of the ethology of jumping spiders (Araneae, Salticidae)', *Bulletin of the British Arachnological Society*, 9(2), pp. 33–7.

Millipedes

Fusco, G., 'Trunk segment numbers and sequential segmentation in myriapods', *Evolution and Development*, 7(6), 2005, pp. 608–17.

Peckre, L.R., Defolie, C., Kappeler, P.M. and Fichtel, C., 'Potential self-medication using millipede secretions in red-fronted lemurs: combining anointment and ingestion for a joint action against gastrointestinal parasites?', *Primates*, 59(5), 2018, pp. 483–94.

Rettenmeyer, C.W., 'The behavior of millipeds found with Neotropical army ants', *Journal of the Kansas Entomological Society*, 35(4), 1962, pp. 377–84.

Mole salamanders

Bogart, J.P., Bi, K., Fu, J., Noble, D.W. and Niedzwiecki, J., 'Unisexual salamanders (genus *Ambystoma*) present a new reproductive mode for eukaryotes', *Genome*, 50(2), 2007, pp. 119–36.

Denton, R.D., Morales, A.E. and Gibbs, H.L., 'Genome-specific histories of divergence and introgression between an allopolyploid unisexual salamander lineage and two ancestral sexual species', *Evolution*, 72(8), 2018, pp. 1689–700.

Mountain tree shrew

Cao, J., Yang, E.B., Su, J.J., Li, Y. and Chow, P., 'The tree shrews: adjuncts and alternatives to primates as models for biomedical research', *Journal of Medical Primatology*, 32(3), 2003, pp. 123–30.

Clarke, C.M., Bauer, U., Ch'ien, C.L., Tuen, A.A., Rembold, K. and Moran, J.A., 'Tree shrew lavatories: a novel nitrogen sequestration strategy in a tropical pitcher plant, *Biology Letters*, 5(5), 2009, pp. 632–5.

Fan, Y., Luo, R., Su, L.Y., Xiang, Q., Yu, D., Xu, L., Chen, J.Q., Bi, R., Wu, D.D., Zheng, P. and Yao, Y.G., 'Does the genetic feature of the Chinese tree shrew (*Tupaia belangeri chinensis*) support its potential as a viable model for Alzheimer's disease research?', *Journal of Alzheimer's Disease*, 61(3), 2018, pp. 1015–28.

Greenwood, M., Clarke, C., Ch'ien, C.L., Gunsalam, A. and Clarke, R.H., 'A unique resource mutualism between the giant Bornean pitcher plant, *Nepenthes rajah*, and members of a small mammal community', *PLoS One*, 6(6), 2011, p .e21114.

Moran, J.A., Clarke, C., Greenwood, M. and Chin, L., 'Tuning of color contrast signals to visual sensitivity maxima of tree shrews by three Bornean highland *Nepenthes* species. *Plant Signaling and Behavior*', 7(10), 2012, pp. 1267–70.

Mudskippers

Lee, H.J., Martinez, C.A., Hertzberg, K.J., Hamilton, A.L. and Graham, J.B., 'Burrow air phase maintenance and respiration by the mudskipper *Scartelaos histophorus* (Gobiidae:

Oxudercinae)', *Journal of Experimental Biology*, 208(1), 2005, pp. 169–77.
Michel, K.B., Heiss, E., Aerts, P. and Van Wassenbergh, S., 'A fish that uses its hydrodynamic tongue to feed on land', *Proceedings of the Royal Society B: Biological Sciences*, 282(1805), 2015, p. 20150057.
Murdy, E.O., 'A taxonomic revision and cladistic analysis of the oxudercine gobies (Gobiidae: Oxudercinae)', *Records of the Australian Museum, Supplement*, *11*, 1989, pp. 1–93.
Sayer, M.D., 'Adaptations of amphibious fish for surviving life out of water', *Fish and Fisheries*, 6(3), 2005, pp. 186–211.

Naked mole-rat

Braude, S., Holtze, S., Begall, S., Brenmoehl, J., Burda, H., Dammann, P., Del Marmol, D., Gorshkova, E., Henning, Y., Hoeflich, A. and Höhn, A., 'Surprisingly long survival of premature conclusions about naked mole-rat biology', *Biological Reviews*, 96(2), 2021, pp. 376–93.
Lewis, K.N. and Buffenstein, R., 'The naked mole-rat: a resilient rodent model of aging, longevity, and healthspan' in *Handbook of the Biology of Aging*, Academic Press, 2016, pp. 179–204).
Ruby, J.G., Smith, M. and Buffenstein, R., 'Naked mole-rat mortality rates defy Gompertzian laws by not increasing with age', *elife*, *7*, 2018, p. e31157.
Watarai, A., Arai, N., Miyawaki, S., Okano, H., Miura, K., Mogi, K. and Kikusui, T., 'Responses to pup vocalizations in subordinate naked mole-rats are induced by estradiol ingested through coprophagy of queen's feces', *Proceedings of the National Academy of Sciences*, 115(37), 2018, pp. 9264–9.

Pangolins

Davit-Béal, T., Tucker, A.S. and Sire, J.Y., 'Loss of teeth and enamel in tetrapods: fossil record, genetic data and morphological adaptations', *Journal of Anatomy*, 214(4), 2009, pp. 477–501.
Emogor, C.A., Ingram, D.J., Coad, L., Worthington, T.A., Dunn, A., Imong, I. and Balmford, A., 'The scale of Nigeria's involvement in the trans-national illegal pangolin trade: temporal and spatial patterns and the effectiveness of wildlife trade regulations', *Biological Conservation*, 264, 2021, p. 109365.

Pseudoscorpion

Del-Claro, K. and Tizo-Pedroso, E., 'Ecological and evolutionary pathways of social behavior in Pseudoscorpions (Arachnida: Pseudoscorpiones)', *Acta Ethologica*, 12(1), 2009, pp. 13–22.
Tizo-Pedroso, E. and Del-Claro, K., 'Matriphagy in the neotropical pseudoscorpion *Paratemnoides nidificator* (Balzan 1888) (Atemnidae)', *The Journal of Arachnology*, 33(3), 2005, pp. 873–7.
Tizo-Pedroso, E. and Del-Claro, K., 'Cooperation in the neotropical pseudoscorpion, *Paratemnoides nidificator* (Balzan, 1888): feeding and dispersal behavior', *Insectes Sociaux*, 54(2), 2007, pp. 124–31.
Tizo-Pedroso, E. and Del-Claro, K., 'Capture of large prey and feeding priority in the cooperative pseudoscorpion.*Paratemnoides nidificator*', *Acta Ethologica*, 21(2), 2018, pp. 109–17.

Red-eyed tree frog

Caldwell, M.S., McDaniel, J.G. and Warkentin, K.M., 'Is it safe? Red-eyed treefrog embryos assessing predation risk use two features of rain vibrations to avoid false alarms', *Animal Behaviour*, 79(2), 2010, pp. 255–60.
Robertson, J.M. and Greene, H.W., 'Bright colour patterns as social signals in nocturnal frogs', *Biological Journal of the Linnean Society*, 121(4), 2017, pp. 849–57.

Saharan silver ant

Pfeffer, S.E., Wahl, V.L., Wittlinger, M. and Wolf, H., 'High-speed locomotion in the Saharan silver ant, *Cataglyphis bombycina*', *Journal of*

Experimental Biology, 222(20), 2019, p. jeb198705.
Shi, N.N., Tsai, C.C., Camino, F., Bernard, G.D., Yu, N. and Wehner, R., 'Keeping cool: enhanced optical reflection and radiative heat dissipation in Saharan silver ants', *Science*, 349(6245), 2015, pp. 298–301.
Wehner, R., Marsh, A.C. and Wehner, S., 'Desert ants on a thermal tightrope', *Nature*, 357(6379), 1992, pp. 586–7.

Saiga antelope

Bekenov, A.B., Grachev, I.A. and Milner-Gulland, E.J., 'The ecology and management of the saiga antelope in Kazakhstan', *Mammal Review*, 28(1), 1998, pp. 1–52.
Kock, R.A., Orynbayev, M., Robinson, S., Zuther, S., Singh, N.J., Beauvais, W., Morgan, E.R., Kerimbayev, A., Khomenko, S., Martineau, H.M. and Rystaeva, R., 'Saigas on the brink: multidisciplinary analysis of the factors influencing mass mortality events', *Science Advances*, 4(1), 2018, p. eaao2314.
Kühl, A., Mysterud, A., Erdnenov, G.I., Lushchekina, A.A., Grachev, I.A., Bekenov, A.B. and Milner-Gulland, E.J., 'The "big spenders" of the steppe: sex-specific maternal allocation and twinning in the saiga antelope', *Proceedings of the Royal Society B: Biological Sciences*, 274(1615), 2007, pp. 1293–9.
Milner-Gulland, E.J., Bukreeva, O.M., Coulson, T., Lushchekina, A.A., Kholodova, M.V., Bekenov, A.B. and Grachev, I.A., 'Reproductive collapse in saiga antelope harems', *Nature*, 422(6928), 2003, p. 135.

Slave-making ant

Achenbach, A. and Foitzik, S., 'First evidence for slave rebellion: enslaved ant workers systematically kill the brood of their social parasite *Protomognathus americanus*', *Evolution: International Journal of Organic Evolution*, 63(4), 2009, pp. 1068–75.
Foitzik, S. and Herbers, J.M., 'Colony structure of a slavemaking ant. II. Frequency of slave raids and impact on the host population', *Evolution*, 55(2), 2001, pp. 316–23.
Pohl, S. and Foitzik, S., 'Slave-making ants prefer larger, better defended host colonies', *Animal Behaviour*, 81(1), 2011, pp. 61–8.

Slow lorises

Nekaris, K.A.I., 'Extreme primates: ecology and evolution of Asian lorises', *Evolutionary Anthropology: Issues, News, and Reviews*, 23(5), 2014, pp. 177–87.
Nekaris, K.A.I., Campbell, N., Coggins, T.G., Rode, E.J. and Nijman, V., 'Tickled to death: analysing public perceptions of "cute" videos of threatened species (slow lorises –*Nycticebus* spp.) on Web 2.0 Sites', *PloS One*, 8(7), 2013, p. e69215.
Nekaris, K., Moore, R.S., Rode, E.J. and Fry, B.G., 'Mad, bad and dangerous to know: the biochemistry, ecology and evolution of slow loris venom', *Journal of Venomous Animals and Toxins including Tropical Diseases*, 19, 2013, pp. 1–10.

Southern grasshopper mouse

Hafner, M.S. and Hafner, D.J., 'Vocalizations of grasshopper mice (genus *Onychomys*)', *Journal of Mammalogy*, 60(1), 1979, pp. 85–94.
McCarty, R., '*Onychomys torridus*', *Mammalian Species*, 59, 1975, pp. 1–5.
McCarty, R. and Southwick, C.H., 'Patterns of parental care in two cricetid rodents, *Onychomys torridus* and *Peromyscus leucopus*', *Animal Behaviour*, 25(4), 1977, pp. 945–8.
Rowe, A.H., Xiao, Y., Rowe, M.P., Cummins, T.R. and Zakon, H.H., 'Voltage-gated sodium channel in grasshopper mice defends against bark scorpion toxin', *Science*, 342(6157), 2013, pp. 441–6.

Tarantulas

Hénaut, Y. and Machkour-M'Rabet, S., 'Predation and Other Interactions' in *New World Tarantulas*, Springer, Cham, 2020, pp. 237–69.
Von May, R., Biggi, E., Cárdenas, H., Diaz, M.I., Alarcón, C., Herrera, V.,

Santa-Cruz, R., Tomasinelli, F., Westeen, E.P., Sánchez-Paredes, C.M., Larson, J.G., Title, P.O., Grundler, M.R., Grundler, M.C., Rabosky, A.R.D., Rabosky, D.L., 'Ecological interactions between arthropods and small vertebrates in a lowland Amazon rainforest', *Amphibian and Reptile Conservation*, 13(1), 2019, pp. 65–77.

Tetradonematid nematode

Poinar, G., 'Nematode parasites and associates of ants: past and present', *Psyche, 2012*(2), 2012.

Poinar, G. and Yanoviak, S.P., '*Myrmeconema neotropicum* n.g., n. sp., a new tetradonematid nematode parasitising South American populations of *Cephalotes atratus* (Hymenoptera: Formicidae), with the discovery of an apparent parasite-induced host morph', *Systematic Parasitology*, 69(2), 2008, pp.145–53.

Yanoviak, S.P., Kaspari, M., Dudley, R. and Poinar Jr, G., 'Parasite-induced fruit mimicry in a tropical canopy ant', *The American Naturalist*, 171(4), 2008, pp. 536–44.

Texas horned lizard

Cooper Jr, W.E. and Sherbrooke, W.C., 'Plesiomorphic escape decisions in cryptic horned lizards (*Phrynosoma*) having highly derived antipredatory defenses', *Ethology*, 116(10), 2010, pp. 920–28.

Holte, A.E. and Houck, M.A., 'Juvenile greater roadrunner (Cuculidae) killed by choking on a Texas horned lizard (Phrynosomatidae)', *The Southwestern Naturalist*, 45(1), 2000, pp. 74–6.

Sherbrooke, W.C. and Middendorf III, G.A., 'Responses of kit foxes (*Vulpes macrotis*) to antipredator blood-squirting and blood of Texas horned lizards (*Phrynosoma cornutum*)', *Copeia*, 2004(3), 2004, pp. 652–8.

Sherbrooke, W.C., 'Rain-harvesting in the lizard, *Phrynosoma cornutum*: behavior and integumental morphology', *Journal of Herpetology*, 24(3), 1900, pp. 302–8.

Velvet worms

Read, V.S.J. and Hughes, R.N., 'Feeding behaviour and prey choice in *Macroperipatus torquatus* (Onychophora)', *Proceedings of the Royal Society of London. Series B: Biological Sciences*, 230(1261), 1987, pp. 483–506.

Reinhard, J. and Rowell, D.M., 'Social behaviour in an Australian velvet worm, *Euperipatoides rowelli* (Onychophora: Peripatopsidae)', *Journal of Zoology*, 267(1), 2005, pp. 1–7.

Tait, N.N. and Briscoe, D.A., 'Sexual head structures in the Onychophora: unique modifications for sperm transfer', *Journal of Natural History*, 24(6), 1990, pp. 1517–27.

Wombats

Triggs, B., ed., *Wombats*, CSIRO Publishing, 2009.

Yang, P.J., Chan, M., Carver, S. and Hu, D.L., 'How do wombats make cubed poo?' in *71st Annual Meeting of the APS Division of Fluid Dynamics* (Vol. 63), 2018.

Yang, P.J., Lee, A.B., Chan, M., Kowalski, M., Qiu, K., Waid, C., Cervantes, G., Magondu, B., Biagioni, M., Vogelnest, L. and Martin, A., 'Intestines of non-uniform stiffness mold the corners of wombat feces', *Soft Matter*, 17(3), 2021, pp. 475–88.

Wood frog

Costanzo, J.P., do Amaral, M.C.F., Rosendale, A.J. and Lee Jr, R.E., 'Hibernation physiology, freezing adaptation and extreme freeze tolerance in a northern population of the wood frog', *Journal of Experimental Biology*, 216(18), 2013, pp. 3461–73.

Jefferson, D.M., Hobson, K.A., Demuth, B.S., Ferrari, M.C. and Chivers, D.P., 'Frugal cannibals: how consuming conspecific tissues can provide conditional benefits to wood frog tadpoles (*Lithobates sylvaticus*). *Naturwissenschaften*', 101(4), 2014, pp. 291–303.

Larson, D.J., Middle, L., Vu, H., Zhang, W., Serianni, A.S., Duman, J. and

Barnes, B.M., 'Wood frog adaptations to overwintering in Alaska: new limits to freezing tolerance', *Journal of Experimental Biology*, 217(12), 2014, pp. 2193–200.

WATER

Amazon river dolphin

Best, R.C. and Da Silva, V.M., '*Inia geoffrensis*', *Mammalian species*, 426, 1993, pp. 1–8.

Da Silva, V., Trujillo, F., Martin, A., Zerbini, A.N., Crespo, E., Aliaga-Rossel, E. and Reeves, R., '*Inia geoffrensis*', *The IUCN Red List of Threatened Species, 2018*, 2018, p. e-T10831A50358152.

Martin, A.R. and Da Silva, V.M.F., 'Sexual dimorphism and body scarring in the boto (Amazon river dolphin) *Inia geoffrensis*', *Marine Mammal Science*, 22(1), 2006, pp. 25–33.

Renjun, L., Gewalt, W., Neurohr, B. and Winkler, A., 'Comparative studies on the behaviour of *Inia geoffrensis* and *Lipotes vexillifer* in artificial environments', *Aquatic Mammals*, 20(1), 1994, pp. 39–45.

Atlantic horseshoe crab

Battelle, B.A., 'The eyes of *Limulus polyphemus* (Xiphosura, Chelicerata) and their afferent and efferent projections', *Arthropod Structure and Development*, 35(4), 2006, pp. 261–74.

Bicknell, R.D. and Pates, S., 'Pictorial atlas of fossil and extant horseshoe crabs, with focus on Xiphosurida', *Frontiers in Earth Science*, 8, 2020, p. 98.

Krisfalusi-Gannon, J., Ali, W., Dellinger, K., Robertson, L., Brady, T.E., Goddard, M.K., Tinker-Kulberg, R., Kepley, C.L. and Dellinger, A.L., 'The role of horseshoe crabs in the biomedical industry and recent trends impacting species sustainability', *Frontiers in Marine Science*, 5, 2018, p. 185.

Bluestreak cleaner wrasse

Abbott, A., 'Animal behaviour: inside the cunning, caring and greedy minds of fish', *Nature News*, 521(7553), 2015, p. 412.

Bshary, R., 'Biting cleaner fish use altruism to deceive image-scoring client reef fish', *Proceedings of the Royal Society of London. Series B: Biological Sciences*, 269(1505), 2002, pp. 2087–93.

Salwiczek, L.H., Prétôt, L., Demarta, L., Proctor, D., Essler, J., Pinto, A.I., Wismer, S., Stoinski, T., Brosnan, S.F. and Bshary, R., 'Adult cleaner wrasse outperform capuchin monkeys, chimpanzees and orang-utans in a complex foraging task derived from cleaner–client reef fish cooperation', *PLoS One*, 7(11), 2012, p. e49068.

Wickler, W., 'Mimicry in tropical fishes', *Philosophical Transactions of the Royal Society of London. Series B: Biological Sciences*, 251(772), 1966, pp. 473–4.

Bobbit worm

Lachat, J. and Haag-Wackernagel, D., 'Novel mobbing strategies of a fish population against a sessile annelid predator', *Scientific Reports*, 6(1), 2016, pp. 1–8.

Uchida, H.O., Tanase, H. and Kubota, S., 'An extraordinarily large specimen of the polychaete worm *Eunice aphroditois* (Pallas) (Order Eunicea) from Shirahama, Wakayama, central Japan', *Kuroshio Biosphere*, 5, 2009, pp. 9–15.

Deep-sea anglerfish

Pietsch, T.W., 'Dimorphism, parasitism, and sex revisited: modes of reproduction among deep-sea ceratioid anglerfishes (Teleostei: Lophiiformes)', *Ichthyological Research*, 52(3), 2005, pp. 207–36.

Swann, J.B., Holland, S.J., Petersen, M., Pietsch, T.W. and Boehm, T., 'The immunogenetics of sexual parasitism', *Science*, 369(6511), 2020, pp. 1608–15.

Ducks

Brennan, P.L., Clark, C.J. and Prum, R.O., 'Explosive eversion and functional morphology of the duck penis supports sexual conflict in waterfowl genitalia', *Proceedings of the Royal*

Society B: Biological Sciences, 277(1686), 2010, pp. 1309–14.

Lyon, B.E. and Eadie, J.M., 'Patterns of host use by a precocial obligate brood parasite, the Black-headed Duck: ecological and evolutionary considerations', *Chinese Birds*, 4(1), 2013, pp. 71–85.

Moeliker, C.W., 'The first case of homosexual necrophilia in the mallard *Anas platyrhynchos* (Aves: Anatidae)', *Deinsea*, 8(1), 2001, pp. 243–8.

Simberloff, D., 'Hybridization between native and introduced wildlife species: importance for conservation', *Wildlife Biology*, 2(3), 1996, pp. 143–50.

Ten Cate, C. and Fullagar, P.J., 'Vocal imitations and production learning by Australian musk ducks (*Biziura lobata*)', *Philosophical Transactions of the Royal Society B: Biological Sciences*, 376(1836), 2021, p. 20200243.

Flukes

Helluy, S. and Thomas, F., 'Parasitic manipulation and neuroinflammation: evidence from the system *Microphallus papillorobustus* (Trematoda) - *Gammarus* (Crustacea)', *Parasites and Vectors*, 3(1), 2010, pp. 1–11.

Levri, E.P. and Fisher, L.M., 'The effect of a trematode parasite (*Microphallus* sp.) on the response of the freshwater snail *Potamopyrgus antipodarum* to light and gravity', *Behaviour*, 137(9), 2000, pp. 1141–52.

McCarthy, H.O., Fitzpatrick, S. and Irwin, S.W.B., 'A transmissible trematode affects the direction and rhythm of movement in a marine gastropod', *Animal Behaviour*, 59(6), 2000, pp. 1161–6.

McCarthy, H.O., Fitzpatrick, S.M. and Irwin, S.W.B., 'Parasite alteration of host shape: a quantitative approach to gigantism helps elucidate evolutionary advantages', *Parasitology*, 2004, 128(1), pp. 7–14.

Gharial

Lang, J.W. and Andrews, H.V., 'Temperature-dependent sex determination in crocodilians', *Journal of Experimental Zoology*, 270(1), 1994, pp. 28–44.

Lang, J.W., 'Adult–young associations of free-living gharials in Chambal River, North India' in *Crocodiles. Proceedings of the 20th Working Meeting of the Crocodile Specialist Group, IUCN, Gland, Switzerland and Cambridge UK*, 2010, p. 142.

Whitaker, R., 'The gharial: going extinct again', *Iguana*, 14(1), 2007, pp. 25–32.

Giant Australian cuttlefish

Brown, C., Garwood, M.P. and Williamson, J.E., 'It pays to cheat: tactical deception in a cephalopod social signalling system', *Biology Letters*, 8(5), 2012, pp. 729–32.

Hall, K. and Hanlon, R., 'Principal features of the mating system of a large spawning aggregation of the giant Australian cuttlefish *Sepia apama* (Mollusca: Cephalopoda)', *Marine Biology*, 140(3), 2002, pp. 533–45.

Hanlon, R.T., Naud, M.J., Shaw, P.W. and Havenhand, J.N., 'Transient sexual mimicry leads to fertilization', *Nature*, 433(7023), 2005, p. 212.

Giant water bug

Ichikawa, N., 'Male counterstrategy against infanticide of the female giant water bug *Lethocerus deyrollei* (Hemiptera: Belostomatidae)', *Journal of Insect Behavior*, 8(2), 1994, pp. 181–8.

Ohba, S.Y., 'Ecology of giant water bugs (Hemiptera: Heteroptera: Belostomatidae)', *Entomological Science*, 22(1), 2019, pp. 6–20.

Greenland shark

MacNeil, M.A., McMeans, B.C., Hussey, N.E., Vecsei, P., Svavarsson, J., Kovacs, K.M., Lydersen, C., Treble, M.A., Skomal, G.B., Ramsey, M. and Fisk, A.T., 'Biology of the Greenland shark *Somniosus microcephalus*', *Journal of Fish Biology*, 80(5), 2012, pp. 991–1018.

Nielsen, J., Hedeholm, R.B., Heinemeier, J., Bushnell, P.G., Christiansen,

J.S., Olsen, J., Ramsey, C.B., Brill, R.W., Simon, M., Steffensen, K.F. and Steffensen, J.F., 'Eye lens radiocarbon reveals centuries of longevity in the Greenland shark (*Somniosus microcephalus*)', *Science*, 353(6300), 2016, pp. 702–4.

Watanabe, Y.Y., Lydersen, C., Fisk, A.T. and Kovacs, K.M., 'The slowest fish: swim speed and tail-beat frequency of Greenland sharks', *Journal of Experimental Marine Biology and Ecology*, 426, 2012, pp. 5–11.

Hagfish

Chaudhary, G., Ewoldt, R.H. and Thiffeault, J.L., 'Unravelling hagfish slime', *Journal of the Royal Society Interface*, 16(150), 2019, p. 20180710.

Haney, W.A., Clark, A.J. and Uyeno, T.A., 'Characterization of body knotting behavior used for escape in a diversity of hagfishes', *Journal of Zoology*, 310(4), 2020, pp. 261–72.

Zintzen, V., Roberts, C.D., Anderson, M.J., Stewart, A.L., Struthers, C.D. and Harvey, E.S., 'Hagfish predatory behaviour and slime defence mechanism', *Scientific Reports*, 1(1), 2011, pp. 1–6.

Harp sponge

Hestetun, J.T., Rapp, H.T. and Pomponi, S., 'Deep-sea carnivorous sponges from the Mariana Islands', *Frontiers in Marine Science*, 6, 2019, p. 371.

Lee, W.L., Reiswig, H.M., Austin, W.C. and Lundsten, L., 'An extraordinary new carnivorous sponge, *Chondrocladia lyra*, in the new subgenus *Symmetrocladia* (Demospongiae, Cladorhizidae), from off of northern California, USA', *Invertebrate Biology*, 131(4), 2012, pp. 259–84.

Herrings

Hunt, K., *Herring: A Global History*, Reaktion Books, 2017.

Mann, D.A., Popper, A.N. and Wilson, B., 'Pacific herring hearing does not include ultrasound', *Biology Letters*, 1(2), 2005, pp. 158–61.

Wahlberg, M. and Westerberg, H., 'Sounds produced by herring (*Clupea harengus*) bubble release', *Aquatic Living Resources*, 16(3), 2003, pp. 271–5.

Wilson, B., Batty, R.S. and Dill, L.M., 'Pacific and Atlantic herring produce burst pulse sounds', *Proceedings of the Royal Society of London. Series B: Biological Sciences*, 271(Suppl. 3), 2004, pp. S95–S97.

Immortal jellyfish

Piraino, S., Boero, F., Aeschbach, B. and Schmid, V., 'Reversing the life cycle: medusae transforming into polyps and cell transdifferentiation in *Turritopsis nutricula* (Cnidaria, Hydrozoa)', *The Biological Bulletin*, 190(3), 1996, pp. 302–12.

Marine iguana

Berger, S., Wikelski, M., Romero, L.M., Kalko, E.K. and Rödl, T., 'Behavioral and physiological adjustments to new predators in an endemic island species, the Galápagos marine iguana', *Hormones and Behavior*, 52(5), 2007, pp. 653–63.

Darwin, C., *Geological observations on South America: Being the third part of the geology of the voyage of the* Beagle, *under the command of Capt. Fitzroy, RN, during the years 1832 to 1836*, Smith, Elder and Co., 1846.

Kruuk, H. and Snell, H., 'Prey selection by feral dogs from a population of marine iguanas (*Amblyrhynchus cristatus*)', *Journal of Applied Ecology*, 1981, pp. 197–204.

Vitousek, M.N., Rubenstein, D.R., Wikelski, M., Reilly, S., McBrayer, L. and Miles, D., 'The evolution of foraging behavior in the Galápagos marine iguana: natural and sexual selection on body size drives ecological, morphological, and behavioral specialization', *Lizard Ecology: The Evolutionary Consequences of Foraging Mode*, 2007, pp. 491–507.

Wikelski, M., 'Evolution of body size in Galápagos marine iguanas', *Proceedings of the Royal Society B: Biological Sciences*, 272(1576), 2005, pp. 1985–93.

Mary River turtle

Campbell, M.A., Connell, M.J., Collett, S.J., Udyawer, V., Crewe, T.L., McDougall, A. and Campbell, H.A., 'The efficacy of protecting turtle nests as a conservation strategy to reverse population decline', *Biological Conservation*, 251, 2020, p. 108769.

Cann, J. and Legler, J.M., 'The Mary River tortoise: a new genus and species of short-necked chelid from Queensland, Australia (Testudines: Pleurodira)', *Chelonian Conservation and Biology*, 1(2), 1994, pp. 81–96.

Clark, N.J., Gordos, M.A. and Franklin, C.E., 'Thermal plasticity of diving behavior, aquatic respiration, and locomotor performance in the Mary River turtle *Elusor macrurus*', *Physiological and Biochemical Zoology*, 81(3), 2008, pp. 301–9.

FitzGibbon, S. and Franklin, C., 'The importance of the cloacal bursae as the primary site of aquatic respiration in the freshwater turtle, *Elseya albagula*', *Australian Zoologist*, 35(2), 2010, pp. 276–82.

Le, M., Reid, B.N., McCord, W.P., Naro-Maciel, E., Raxworthy, C.J., Amato, G. and Georges, A., 'Resolving the phylogenetic history of the short-necked turtles, genera *Elseya* and *Myuchelys* (Testudines: Chelidae) from Australia and New Guinea', *Molecular Phylogenetics and Evolution*, 68(2), 2013, pp. 251–8.

Mimic octopus

Hanlon, R.T., Conroy, L.A. and Forsythe, J.W., 'Mimicry and foraging behaviour of two tropical sand-flat octopus species off North Sulawesi, Indonesia', *Biological Journal of the Linnean Society*, 93(1), 2008, pp. 23–38.

Norman, M.D., Finn, J. and Tregenza, T., 'Dynamic mimicry in an Indo-Malayan octopus', *Proceedings of the Royal Society of London. Series B: Biological Sciences*, 268(1478), 2001, pp. 1755–8.

Olm

Balázs, G., Lewarne, B. and Herczeg, G., 'Extreme site fidelity of the olm (*Proteus anguinus*) revealed by a long-term capture–mark–recapture study', *Journal of Zoology*, 311(2), 2020, pp. 99–105.

Culver, D.C. and White, W.B., *Encyclopedia of Caves* (2nd edn), Waltham, MA: Elsevier/Academic Press, 2012.

Voituron, Y., de Fraipont, M., Issartel, J., Guillaume, O. and Clobert, J., 'Extreme lifespan of the human fish (*Proteus anguinus*): a challenge for ageing mechanisms', *Biology Letters*, 7(1), 2011, pp. 105–7.

Peacock mantis shrimp

Daly, I.M., How, M.J., Partridge, J.C., Temple, S.E., Marshall, N.J., Cronin, T.W. and Roberts, N.W., 'Dynamic polarization vision in mantis shrimps', *Nature Communications*, 7(1), 2016, pp. 1–9.

Patek, S.A. and Caldwell, R.L., 'Extreme impact and cavitation forces of a biological hammer: strike forces of the peacock mantis shrimp *Odontodactylus scyllarus*', *Journal of Experimental Biology*, 208(19), 2005, pp. 3655–64.

Vetter, K.M. and Caldwell, R.L., 'Individual recognition in stomatopods' in *Social Recognition in Invertebrates*, Springer, Cham, 2015, pp. 17–36.

Weaver, J.C., Milliron, G.W., Miserez, A., Evans-Lutterodt, K., Herrera, S., Gallana, I., Mershon, W.J., Swanson, B., Zavattieri, P., DiMasi, E. and Kisailus, D., 'The stomatopod dactyl club: a formidable damage-tolerant biological hammer', *Science*, 336(6086), 2012, pp. 1275–80.

Pearlfish

Lagardère, J.P., Millot, S. and Parmentier, E., 'Aspects of sound communication in the pearlfish *Carapus boraborensis* and *Carapus homei* (Carapidae)', *Journal of Experimental Zoology, Part A: Comparative Experimental Biology*, 303(12), 2005, pp. 1066–74.

Meyer-Rochow, V.B., 'Comparison between 15 *Carapus mourlani* in a single holothurian and 19 *C. mourlani* from starfish', *Copeia*, 1977(3), 1977, pp. 582–4.

Parmentier, E. and Vandewalle, P., 'Further insight on carapid–holothuroid relationships', *Marine Biology*, 146(3), 2005, pp. 455–65.

Trott, L.B., 'A general review of the pearlfishes (Pisces, Carapidae)', *Bulletin of Marine Science*, 31(3), 1981, pp. 623–9.

***Piure* sea squirt**

Lagger, C., Häussermann, V., Försterra, G. and Tatián, M., 'Ascidians from the southern Chilean Comau Fjord', *Spixiana*, 32(2), 2009, pp. 173–85.

Lambert, G., Karney, R.C., Rhee, W.Y. and Carman, M.R., 'Wild and cultured edible tunicates: a review', *Management of Biological Invasions*, 7(1), 2016, pp. 59–66.

Manríquez, P.H. and Castilla, J.C., 'Self-fertilization as an alternative mode of reproduction in the solitary tunicate *Pyura chilensis*', *Marine Ecology Progress Series*, 305, 2005, pp. 113–125.

Segovia, N.I., González-Wevar, C.A. and Haye, P.A., 'Signatures of local adaptation in the spatial genetic structure of the ascidian *Pyura chilensis* along the southeast Pacific coast', *Scientific Reports*, 10(1), 2020, pp. 1–14.

Platypus

Grützner, F., Nixon, B. and Jones, R.C., 'Reproductive biology in egg-laying mammals', *Sexual Development*, 2(3), 2008, pp. 115–27.

Newman, J., Sharp, J.A., Enjapoori, A.K., Bentley, J., Nicholas, K.R., Adams, T.E. and Peat, T.S., 'Structural characterization of a novel monotreme-specific protein with antimicrobial activity from the milk of the platypus', *Acta Crystallographica Section F: Structural Biology Communications*, 74(1), 2018, pp. 39–45.

Warren, W.C., Hillier, L.W., Graves, J.A.M., Birney, E., Ponting, C.P., Grützner, F., Belov, K., Miller, W., Clarke, L., Chinwalla, A.T. and Yang, S.P., 'Genome analysis of the platypus reveals unique signatures of evolution', *Nature*, 453(7192), 2008, p. 175.

Racing stripe flatworm

Michiels, N.K. and Newman, L.J., 'Sex and violence in hermaphrodites', *Nature*, 391(6668), 1998, p. 647.

Ramm, S.A., Schlatter, A., Poirier, M. and Schärer, L., 'Hypodermic self-insemination as a reproductive assurance strategy', *Proceedings of the Royal Society B: Biological Sciences*, 282(1811), 2015, p. 20150660.

Roving coral grouper

Bshary, R., Hohner, A., Ait-el-Djoudi, K. and Fricke, H., 'Interspecific communicative and coordinated hunting between groupers and giant moray eels in the Red Sea', *PLoS Biology*, 4(12), 2006, p. e431.

Sampaio, E., Seco, M.C., Rosa, R. and Gingins, S., 'Octopuses punch fishes during collaborative interspecific hunting events', *Ecology*, 2020, p. e03266.

Vail, A.L., Manica, A. and Bshary, R., 'Referential gestures in fish collaborative hunting', *Nature Communications*, 4(1), 2013, pp. 1–7.

Sacoglossan sea slugs

Mitoh, S. and Yusa, Y., 'Extreme autotomy and whole-body regeneration in photosynthetic sea slugs', *Current Biology*, 31(5), 2021, pp. R233–R234.

Shiroyama, H., Mitoh, S., Ida, T.Y. and Yusa, Y., 'Adaptive significance of light and food for a kleptoplastic sea slug: implications for photosynthesis', *Oecologia*, 194(3), 2020, pp. 455–63.

Wägele, H., 'Photosynthesis and the role of plastids (kleptoplastids) in Sacoglossa (Heterobranchia, Gastropoda): a short review', *Aquatic Science and Management*, 3(1), 2015, pp. 1–7.

Sea cucumbers

Fabinyi, M., Barclay, K. and Eriksson, H., 'Chinese trader perceptions on sourcing and consumption of endangered seafood', *Frontiers in Marine Science*, 4, 2017, p. 181.

Laxminarayana, A., 'Asexual reproduction by induced transverse

fission in the sea cucumbers *Bohadschia marmorata* and *Holothuria atra*', *SPC Beche-de-Mer Information Bulletin*, 23, 2006, pp. 35–7.
Motokawa, T. and Tsuchi, A., 'Dynamic mechanical properties of body-wall dermis in various mechanical states and their implications for the behavior of sea cucumbers', *The Biological Bulletin*, 205(3), 2003, pp. 261–75.
Purcell, S.W., Conand, C., Uthicke, S. and Byrne, M., 'Ecological roles of exploited sea cucumbers', *Oceanography and Marine Biology: An Annual Review*, 54, 2016, pp. 367–86.

Sea walnut
Schofield, P.J. and Brown, M.E., 'Invasive species: ocean ecosystem case studies for earth systems and environmental sciences', *Reference Module in Earth Systems and Environmental Sciences*, 2016.
Tamm, S.L., 'Defecation by the ctenophore *Mnemiopsis leidyi* occurs with an ultradian rhythm through a single transient anal pore', *Invertebrate Biology*, 138(1), 2019, pp. 3–16.

Spiny dye murex
Lahbib, Y., Abidli, S. and Trigui-El Menif, N., 'First assessment of the effectiveness of the international convention on the control of harmful anti-fouling systems on ships in Tunisia using imposex in *Hexaplex trunculus* as biomarker', *Marine Pollution Bulletin*, 128, 2018, pp. 17–23.
Vasconcelos, P., Moura, P., Barroso, C.M. and Gaspar, M.B., 'Size matters: importance of penis length variation on reproduction studies and imposex monitoring in *Bolinus brandaris* (Gastropoda: Muricidae)', *Hydrobiologia*, 661(1), 2011, pp. 363–75.

Surinam toad
Fernandez, E., Irish, F. and Cundall, D., 'How a frog, *Pipa pipa*, succeeds or fails in catching fish', *Copeia*, 105(1), 2017, pp. 108–19.
Rabb, G.B. and Rabb, M.S., 'On the mating and egg-laying behavior of the Surinam toad, *Pipa pipa*', *Copeia*, 1960(4), 1960, pp. 271–6.

Tongue-eating louse
Brusca, R.C. and Gilligan, M.R., 'Tongue replacement in a marine fish (*Lutjanus guttatus*) by a parasitic isopod (Crustacea: Isopoda)', *Copeia*, 1983(3), 1983, pp. 813–16.
Parker, D. and Booth, A.J., 'The tongue-replacing isopod *Cymothoa borbonica* reduces the growth of largespot pompano *Trachinotus botla*', *Marine Biology*, 160(11), 2013, pp. 2943–50.
Ruiz, A. and Madrid, J., 'Studies on the biology of the parasitic isopod *Cymothoa exigua* Schioedte and Meinert, 1884, and its relationship with the snapper *Lutjanus peru* (Pisces: Lutjanidae) Nichols and Murphy, 1922, from commercial catch in Michoacan', *Ciencias Marinas*, 18(1), 1992, pp. 19–34.

Water bears
Guidetti, R., Rizzo, A.M., Altiero, T. and Rebecchi, L., 'What can we learn from the toughest animals of the Earth? Water bears (tardigrades) as multicellular model organisms in order to perform scientific preparations for lunar exploration', *Planetary and Space Science*, 74(1), 2012, pp. 97–102.
Traspas, A. and Burchell, M.J.,'Tardigrade survival limits in high-speed impacts – implications for panspermia and collection of samples from plumes emitted by ice worlds', *Astrobiology*, 21(7), 2021, pp. 845–52.

Wattled jacana
Emlen, S.T. and Wrege, P.H., 'Size dimorphism, intrasexual competition, and sexual selection in wattled jacana (*Jacana jacana*), a sex-role-reversed shorebird in Panama', *The Auk*, 121(2), 2004, pp. 391–403.
Emlen, S.T. and Wrege, P.H., 'Division of labour in parental care behaviour of a sex-role-reversed shorebird, the wattled jacana', *Animal Behaviour*, 68(4), 2004, pp. 847–55.

Yeti crab
Marsh, L., Copley, J.T., Tyler, P.A. and Thatje, S., 'In hot and cold water: differential life-history traits are key to success in contrasting thermal deep-sea environments', *Journal of Animal Ecology*, 84(4), 2015, pp. 898–913.
Thatje, S., Marsh, L., Roterman, C.N., Mavrogordato, M.N. and Linse, K., 'Adaptations to hydrothermal vent life in *Kiwa tyleri*, a new species of yeti crab from the East Scotia Ridge, Antarctica', *PLoS One*, 10(6), 2015, p. e0127621.
Thurber, A.R., Jones, W.J. and Schnabel, K., 'Dancing for food in the deep sea: bacterial farming by a new species of yeti crab', *PLoS One*, 6(11), 2011, p. e26243.

Zombie worms
Rouse, G.W., Goffredi, S.K. and Vrijenhoek, R.C., 'Osedax: bone-eating marine worms with dwarf males', *Science*, 305(5684), 2004, pp. 668–71.
Smith, C.R., Glover, A.G., Treude, T., Higgs, N.D. and Amon, D.J., 'Whale-fall ecosystems: recent insights into ecology, paleoecology, and evolution', *Annual Review of Marine Science*, 7, 2015, pp. 571–96.
Tresguerres, M., Katz, S. and Rouse, G.W., 'How to get into bones: proton pump and carbonic anhydrase in *Osedax* boneworms', *Proceedings of the Royal Society B: Biological Sciences*, 280(1761), 2013, p. 20130625.
Vrijenhoek, R.C., Johnson, S.B. and Rouse, G.W., 'Bone-eating *Osedax* females and their "harems" of dwarf males are recruited from a common larval pool', *Molecular Ecology*, 17(20), 2008, pp. 4535–44.

AIR

Bees
Michener, C.D., *The Bees of the World* (Vol. 1), JHU Press, 2000.
Mikát, M., Janošík, L., Černá, K., Matoušková, E., Hadrava, J., Bureš, V. and Straka, J., 'Polyandrous bee provides extended offspring care biparentally as an alternative to monandry-based eusociality', *Proceedings of the National Academy of Sciences*, 116(13), 2019, pp. 6238–43.
Oldroyd, B.P., Yagound, B., Allsopp, M.H., Holmes, M.J., Buchmann, G., Zayed, A. and Beekman, M., 'Adaptive, caste-specific changes to recombination rates in a thelytokous honeybee population', *Proceedings of the Royal Society B: Biological Sciences*, 288(1952), 2021, p. 20210729.

Bombardier beetles
Arndt, E.M., Moore, W., Lee, W.K. and Ortiz, C., 'Mechanistic origins of bombardier beetle (Brachinini) explosion-induced defensive spray pulsation', *Science*, 348(6234), 2015, pp. 563–7.
Eisner, T. and Aneshansley, D.J., 'Spray aiming in the bombardier beetle: photographic evidence', *Proceedings of the National Academy of Sciences*, 96(17), 1999, pp. 9705–9.
Sugiura, S. and Sato, T., 'Successful escape of bombardier beetles from predator digestive systems', *Biology Letters*, 14(2), 2018, p. 20170647.

Boobies
Castillo-Guerrero, J.A., González-Medina, E. and Mellink, E., 'Adoption and infanticide in an altricial colonial seabird, the blue-footed booby: the roles of nest density, breeding success, and sex-biased behavior', *Journal of Ornithology*, 155(1), 2014, pp. 135–44.
Grace, J.K., Dean, K., Ottinger, M.A. and Anderson, D.J., 'Hormonal effects of maltreatment in Nazca booby nestlings: implications for the "cycle of violence"', *Hormones and Behavior*, 60(1), 2011, pp. 78–85.
Lougheed, L.W. and Anderson, D.J., 'Parent blue-footed boobies suppress siblicidal behavior of offspring', *Behavioral Ecology and Sociobiology*, 45(1), 1999, pp. 11–18.
Maness, T.J. and Anderson, D.J., 'Mate rotation by female choice and

coercive divorce in Nazca boobies, *Sula granti*', *Animal Behaviour*, 76(4), 2008, pp. 1267–77.

California scrub-jay

Clayton, N.S., Dally, J.M. and Emery, N.J., 'Social cognition by food-caching corvids. The western scrub-jay as a natural psychologist', *Philosophical Transactions of the Royal Society B: Biological Sciences*, 362(1480), 2007, pp. 507–22.

Wiles, G.J. and McAllister, K.R., 'Records of anting by birds in Washington and Oregon', *Washington Birds*, 11, 2011, pp. 28–34.

Caribbean reef squid

Maciá, S., Robinson, M.P., Craze, P., Dalton, R. and Thomas, J.D., 'New observations on airborne jet propulsion (flight) in squid, with a review of previous reports', *Journal of Molluscan Studies*, 70(3), 2004, pp. 297–9.

Mather, J., 'Mating games squid play: reproductive behaviour and sexual skin displays in Caribbean reef squid *Sepioteuthis sepioidea*', *Marine and Freshwater Behaviour and Physiology*, 49(6), 2016, pp. 359–73.

Chatham Island black robin

Butchart, S.H., Stattersfield, A.J. and Collar, N.J., 'How many bird extinctions have we prevented?', *Oryx*, 40(3), 2006, pp. 266–78.

Massaro, M., Sainudiin, R., Merton, D., Briskie, J.V., Poole, A.M. and Hale, M.L., 'Human-assisted spread of a maladaptive behavior in a critically endangered bird', *PloS One*, 8(12), 2013, p. e79066.

Merton, D.V., 'Cross-fostering of the Chatham Island black robin', *New Zealand Journal of Ecology*, 6, 1983, pp. 156–7.

Merton, D., 'The legacy of "Old Blue"', *New Zealand Journal of Ecology*, 16(2), 1992, pp. 65–8.

Common pigeon

Gagliardo, A., 'Forty years of olfactory navigation in birds', *Journal of Experimental Biology*, 216(12), 2013, pp. 2165–71.

Mosco, R., *A Pocket Guide to Pigeon Watching: Getting to Know the World's Most Misunderstood Bird*, Workman Publishing, 2021.

Sales, J. and Janssens, G.P.J., 'Nutrition of the domestic pigeon (*Columba livia domestica*)', *World's Poultry Science Journal*, 59(2), 2003, pp. 221–32.

Stringham, S.A., Mulroy, E.E., Xing, J., Record, D., Guernsey, M.W., Aldenhoven, J.T., Osborne, E.J. and Shapiro, M.D., 'Divergence, convergence, and the ancestry of feral populations in the domestic rock pigeon', *Current Biology*, 22(4), 2012, pp. 302–8.

Common potoo

Borrero, J.I., 'Notes on the structure of the upper eyelid of potoos (*Nyctibius*)', *The Condor*, 76(2), 1974, pp. 210–11.

Cestari, C., Guaraldo, A.C. and Gussoni, C.O., 'Nestling behavior and parental care of the common potoo (*Nyctibius griseus*) in southeastern Brazil', *The Wilson Journal of Ornithology*, 123(1), 2011, pp. 102–6.

Common swift

Åkesson, S. and Bianco, G., 'Wind-assisted sprint migration in northern swifts', *Iscience*, 24(6), 2021, p. 102474.

Hedenström, A., Norevik, G., Warfvinge, K., Andersson, A., Bäckman, J. and Åkesson, S., 'Annual 10-month aerial life phase in the common swift *Apus apus*', *Current Biology*, 26(22), 2016, pp. 3066–70.

Wright, J., Markman, S. and Denney, S.M., 'Facultative adjustment of pre-fledging mass loss by nestling swifts preparing for flight', *Proceedings of the Royal Society B: Biological Sciences*, 273(1596), 2006, pp. 1895–900.

Common vampire bat

Carter, G.G. and Wilkinson, G.S., 'Food sharing in vampire bats: reciprocal help predicts donations more than relatedness or harassment', *Proceedings of the Royal Society B: Biological Sciences*, 280(1753), 2013, p. 20122573.

Razik, I., Brown, B.K., Page, R.A. and Carter, G.G., 'Non-kin adoption in the common vampire bat', *Royal Society Open Science*, 8(2), 2021, p. 201927.

Wilkinson, G.S., 'Vampire bats', *Current Biology*, 29(23), 2019, pp. R1216–R1217.

Dragonflies

Hobson, K.A., Anderson, R.C., Soto, D.X. and Wassenaar, L.I., 'Isotopic evidence that dragonflies (*Pantala flavescens*) migrating through the Maldives come from the northern Indian subcontinent', *PloS One*, 7(12), 2012, p. e52594.

Khelifa, R., 'Faking death to avoid male coercion: extreme sexual conflict resolution in a dragonfly', *Ecology*, 98(6), 2017, pp. 1724–6.

Olesen, J., 'The hydraulic mechanism of labial extension and jet propulsion in dragonfly nymphs', *Journal of Comparative Physiology*, 81(1), 1972, pp. 53–5.

Emerald cockroach wasp

Fox, E.G.P., Bressan-Nascimento, S. and Eizemberg, R., 'Notes on the biology and behaviour of the jewel wasp, *Ampulex compressa* (Fabricius, 1781) (Hymenoptera; Ampulicidae), in the laboratory, including first record of gregarious reproduction', *Entomological News*, 120(4), 2009, pp. 430–37.

Gal, R., Rosenberg, L.A. and Libersat, F., 'Parasitoid wasp uses a venom cocktail injected into the brain to manipulate the behavior and metabolism of its cockroach prey', *Archives of Insect Biochemistry and Physiology* (published in collaboration with the Entomological Society of America), 60(4), 2005, pp. 198–208.

Flying fish

Byrnes, G. and Spence, A.J., 'Ecological and biomechanical insights into the evolution of gliding in mammals', Integrative and Comparative Biology, 51(6), 2011, pp. 991–1001.

Daane, J.M., Blum, N., Lanni, J., Boldt, H., Iovine, M.K., Higdon, C.W., Johnson, S.L., Lovejoy, N.R. and Harris, M.P., 'Novel regulators of growth identified in the evolution of fin proportion in flying fish', *bioRxiv*, 2021.

Davenport, J., 'How and why do flying fish fly?', *Reviews in Fish Biology and Fisheries*, 4(2), 1994, pp. 184–214.

Guianan cock-of-the-rock

Trail, P.W., 'The courtship behavior of the lek-breeding Guianan cock-of-the-rock: a lek's icon', *American Birds*, 39(3), 1985, pp. 235–40.

Trail, P.W., 'Predation and antipredator behavior at Guianan cock-of-the-rock leks', *The Auk*, 104(3), 1987, pp. 496–507.

Hummingbirds

Clark, C.J., 'Courtship dives of Anna's hummingbird offer insights into flight performance limits', *Proceedings of the Royal Society B: Biological Sciences*, 276(1670), 2009, pp. 3047–52.

Soteras, F., Moré, M., Ibañez, A.C., Iglesias, M.D.R. and Cocucci, A.A., 'Range overlap between the sword-billed hummingbird and its guild of long-flowered species: an approach to the study of a coevolutionary mosaic', *PloS One*, 13(12), 2018, p. e0209742.

Warrick, D., Hedrick, T., Fernández, M.J., Tobalske, B. and Biewener, A., 'Hummingbird flight', *Current Biology*, 22(12), 2012, pp. R472–R477.

Julia butterfly

Benson, W.W., Brown Jr, K.S. and Gilbert, L.E., 'Coevolution of plants and herbivores: passion flower butterflies', *Evolution*, 1975, pp. 659–80.

de Castro, E.C., Zagrobelny, M., Cardoso, M.Z. and Bak, S., 'The arms race between heliconiine butterflies and Passiflora plants – new insights on an ancient subject', *Biological Reviews*, 93(1), 2018, pp. 555–73.

de la Rosa, C.L., 'Additional observations of lachryphagous butterflies and bees', *Frontiers in Ecology and the Environment*, 12(4), 2014, p. 210.

Laysan albatross

Rice, D.W. and Kenyon, K.W., 'Breeding cycles and behavior of Laysan and black-footed albatrosses', *The Auk*, 79(4), 1962, pp. 517–67.

Young, L.C., Zaun, B.J. and VanderWerf, E.A., 'Successful same-sex pairing in Laysan albatross', *Biology Letters*, 4(4), 2008, pp. 323–5.

Marabou stork

Francis, R.J., Kingsford, R.T., Murray-Hudson, M. and Brandis, K.J., 'Urban waste no replacement for natural foods – marabou storks in Botswana', *Journal of Urban Ecology*, 7(1), 2021, p. juab003.

Hancock, J., Kushlan, J.A. and Kahl, M.P., *Storks, Ibises and Spoonbills of the World*, A&C Black, 2010.

Moths

Barber, J.R. and Kawahara, A.Y., 'Hawkmoths produce anti-bat ultrasound', *Biology Letters*, 9(4), 2013, p. 20130161.

Neil, T.R., Shen, Z., Robert, D., Drinkwater, B.W. and Holderied, M.W., 'Moth wings are acoustic metamaterials', *Proceedings of the National Academy of Sciences*, 117(49), 2020, pp. 31134–41.

Rubin, J.J., Hamilton, C.A., McClure, C.J., Chadwell, B.A., Kawahara, A.Y. and Barber, J.R., 'The evolution of anti-bat sensory illusions in moths', *Science Advances*, 4(7), 2018, p. eaar7428.

Ter Hofstede, H.M. and Ratcliffe, J.M., 'Evolutionary escalation: the bat–moth arms race', *Journal of Experimental Biology*, 219(11), 2016, pp. 1589–1602.

New Caledonian crow

Jelbert, S.A., Hosking, R.J., Taylor, A.H. and Gray, R.D., 'Mental template matching is a potential cultural transmission mechanism for New Caledonian crow tool manufacturing traditions', *Scientific Reports*, 8(1), 2018, pp. 1–8.

Kenward, B., Rutz, C., Weir, A.A., Chappell, J. and Kacelnik, A., 'Morphology and sexual dimorphism of the New Caledonian crow *Corvus moneduloides*, with notes on its behaviour and ecology', *Ibis*, 146(4), 2004, pp. 652–60.

Troscianko, J. and Rutz, C., 'Activity profiles and hook-tool use of New Caledonian crows recorded by bird-borne video cameras', *Biology Letters*, 11(12), 2015, p. 20150777.

von Bayern, A.M.P., Danel, S., Auersperg, A.M.I., Mioduszewska, B. and Kacelnik, A., 'Compound tool construction by New Caledonian crows', *Scientific Reports*, 8(1), 2018, pp. 1–8.

Old World fruit bats

Banerjee, A., Baker, M.L., Kulcsar, K., Misra, V., Plowright, R. and Mossman, K., 'Novel insights into immune systems of bats', *Frontiers in Immunology*, 11, 2020, p. 26.

Corlett, R.T., 'Frugivory and seed dispersal by vertebrates in tropical and subtropical Asia: an update', *Global Ecology and Conservation*, 11, 2017, pp.1–22.

Fleming, T.H., Geiselman, C. and Kress, W.J., 'The evolution of bat pollination: a phylogenetic perspective', *Annals of Botany*, 104(6), 2009, pp. 1017–43.

Sugita, N., 'Homosexual fellatio: erect penis licking between male Bonin flying foxes *Pteropus pselaphon*', *PloS One*, 11(11), 2016, p. e0166024.

Orchid mantis

Mizuno, T., Yamaguchi, S., Yamamoto, I., Yamaoka, R. and Akino, T. '"Double-trick" visual and chemical mimicry by the juvenile orchid mantis *Hymenopus coronatus* used in predation of the oriental honeybee *Apis cerana*', *Zoological Science*, 31(12), 2014, pp. 795–801.

O'Hanlon, J.C., Holwell, G.I. and Herberstein, M.E., 'Pollinator deception in the orchid mantis', *The American Naturalist*, 183(1), 2014, pp. 126–32.

O'Hanlon, J.C., 'Orchid mantis', *Current Biology*, 26(4), 2016, pp. R145–R146.

Svenson, G.J., Brannoch, S.K., Rodrigues, H.M., O'Hanlon, J.C. and Wieland, F., 'Selection for predation, not female fecundity, explains sexual size dimorphism in the orchid mantises', *Scientific Reports*, 6(1), 2016, pp. 1–9.

Paradise tree snake

Holden, D., Socha, J.J., Cardwell, N.D. and Vlachos, P.P., 'Aerodynamics of the flying snake *Chrysopelea paradisi*: how a bluff body cross-sectional shape contributes to gliding performance', *Journal of Experimental Biology*, 217(3), 2014, pp. 382–94.

Socha, J.J., 'Gliding flight in *Chrysopelea*: turning a snake into a wing', *Integrative and Comparative Biology*, 51(6), 2011, pp. 969–82.

Periodical cicadas

Cooley, J.R., Marshall, D.C. and Hill, K.B., 'A specialized fungal parasite (*Massospora cicadina*) hijacks the sexual signals of periodical cicadas (Hemiptera: Cicadidae: Magicicada)', *Scientific Reports*, 8(1), 2018, pp. 1–7.

Kritsky, G., 'One for the books: the 2021 emergence of the periodical cicada Brood X', *American Entomologist*, 67(4), 2021, pp. 40–46.

Williams, K.S. and Simon, C., 'The ecology, behavior, and evolution of periodical cicadas', *Annual Review of Entomology*, 40(1), 1995, pp. 269–95.

Regent honeyeater

Crates, R., Langmore, N., Ranjard, L., Stojanovic, D., Rayner, L., Ingwersen, D. and Heinsohn, R., 'Loss of vocal culture and fitness costs in a critically endangered songbird', *Proceedings of the Royal Society B: Biological Sciences*, 288(1947), 2021, p. 20210225.

Crates, R., Rayner, L., Stojanovic, D., Webb, M., Terauds, A. and Heinsohn, R., 'Contemporary breeding biology of critically endangered regent honeyeaters: implications for conservation', *Ibis*, 161(3), 2019, pp. 521–32.

Tripovich, J.S., Popovic, G., Elphinstone, A., Ingwersen, D., Johnson, G., Schmelitschek, E., Wilkin, D., Taylor, G. and Pitcher, B.J., 'Born to be wild: evaluating the zoo-based regent honeyeater breed for release program to optimise individual success and conservation outcomes in the wild', *Frontiers in Conservation Science*, 2, 2021, p. 16.

Sociable weaver

Leighton, G.M. and Meiden, L.V., 'Sociable weavers increase cooperative nest construction after suffering aggression', *PLoS One*, 11(3), 2016, p. e0150953.

Lowney, A.M., Bolopo, D., Krochuk, B.A. and Thomson, R.L., 'The large communal nests of sociable weavers provide year-round insulated refuge for weavers and pygmy falcons', *Frontiers in Ecology and Evolution*, 8, 2020, p. 357.

Lowney, A.M. and Thomson, R.L., 'Ecological engineering across a temporal gradient: sociable weaver colonies create year-round animal biodiversity hotspots', *Journal of Animal Ecology*, 90(10), 2021, pp. 2362–76.

Vampire finch

Bowman, R.I. and Billeb, S.L., 'Blood-eating in a Galápagos finch', *Living Bird*, 4(2), 1965, p. 9.

Michel, A.J., Ward, L.M., Goffredi, S.K., Dawson, K.S., Baldassarre, D.T., Brenner, A., Gotanda, K.M., McCormack, J.E., Mullin, S.W., O'Neill, A. and Tender, G.S., 'The gut of the finch: uniqueness of the gut microbiome of the Galápagos vampire finch', *Microbiome*, 6(1), 2018, pp. 1–14.

Schluter, D. and Grant, P.R., 'Ecological correlates of morphological evolution in a Darwin's finch, *Geospiza difficilis*', *Evolution*, 1984, pp. 856–69.

Tebbich, S., Sterelny, K. and Teschke, I., 'The tale of the finch: adaptive radiation and behavioural flexibility', *Philosophical Transactions of the Royal Society B: Biological Sciences*, 365(1543), 2010, pp. 1099–1109.

White butterfly parasite wasp

Harvey, J.A., Vet, L.E., Witjes, L.M. and Bezemer, T.M., 'Remarkable similarity in body mass of a secondary hyperparasitoid *Lysibia nana* and its primary parasitoid host *Cotesia glomerata* emerging from cocoons of comparable size', *Archives of Insect Biochemistry and Physiology* (published in collaboration with the Entomological Society of America), 61(3), 2006, pp. 170–83.

Zhu, F., Broekgaarden, C., Weldegergis, B.T., Harvey, J.A., Vosman, B., Dicke, M. and Poelman, E.H., 'Parasitism overrides herbivore identity allowing hyperparasitoids to locate their parasitoid host using herbivore-induced plant volatiles', *Molecular Ecology*, 24(11), 2015, pp. 2886–99.

White-rumped vulture

Galligan, T.H., Bhusal, K.P., Paudel, K., Chapagain, D., Joshi, A.B., Chaudhary, I.P., Chaudhary, A., Baral, H.S., Cuthbert, R.J. and Green, R.E., 'Partial recovery of critically endangered *Gyps* vulture populations in Nepal', *Bird Conservation International*, 30(1), 2020, pp. 87–102.

Kanaujia, A. and Kushwaha, S., 'Vulnerable vultures of India: population, ecology and conservation' in *Rare Animals of India*, Bentham Science Publishers, UAE, 2013, pp. 113–144.

Prakash, V., Pain, D.J., Cunningham, A.A., Donald, P.F., Prakash, N., Verma, A., Gargi, R., Sivakumar, S. and Rahmani, A.R., 'Catastrophic collapse of Indian white-backed *Gyps bengalensis* and long-billed *Gyps indicus* vulture populations', *Biological Conservation*, 109(3), 2003, pp. 381–90.

Zebra finch

Boucaud, I.C., Mariette, M.M., Villain, A.S. and Vignal, C., 'Vocal negotiation over parental care? Acoustic communication at the nest predicts partners' incubation share', *Biological Journal of the Linnean Society*, 117(2), 2016, pp. 322–36.

Golüke, S., Bischof, H.J. and Caspers, B.A., 'Nestling odour modulates behavioural response in male, but not in female zebra finches', *Scientific Reports*, 11(1), 2021.

Mariette, M.M. and Buchanan, K.L., 'Prenatal acoustic communication programs offspring for high post-hatching temperatures in a songbird', *Science*, 353(6301), 2016, pp. 812–14.

Schielzeth, H. and Bolund, E., 'Patterns of conspecific brood parasitism in zebra finches', *Animal Behaviour*, 79(6), 2010, pp. 1329–37.

Zann, R.A., *The Zebra Finch: a Synthesis of Field and Laboratory Studies* (Vol. 5), Oxford University Press, 1996.

Acknowledgments

Writing a popular science book is not a one-person job, and *The Modern Bestiary* would not have emerged in its current form if it hadn't been for contributions from colleagues, friends, relations and complete strangers. As such, I am indebted to the following people:

Firstly, to the myriad of researchers whose work I cite—who counted herring farts, swapped albatross eggs for beer cans, documented flying fox blowjobs, built stilts for ants, got aye-ayes tipsy, and did much, much more—to the giants on whose shoulders I can now stand.

Secondly, to all the inspirational, supportive, and infinitely patient zoologists, ecologists, and biologists whom I have reached out to directly, and who proved invaluable at pitching ideas, fact-checking the entries, and recommending reading materials—in particular to Michael Brooke, Amanda Callaghan, Jaime Chaves, Ada Grabowska-Zhang, Chris Hassall, Steph Holt, Jan Kamler, Freya van Kesteren, Silvia Maciá, Louis Phipps, Brian Pickles, Tara Pirie, Lorenzo Santorelli, Deepa Senapathi, Emilia Skirmuntt, John Sumpter, Richard Walters, Arron Watson, and Andrzej Wolniewicz.

Thirdly, to the "normal"—i.e., not zoologist—friends who made a mark on this book. To a very encouraging theologian, Ignacio Silva, for help with bestiaries at large. To Konrad Suder Chatterjee, a classicist, for all the help with translating the often-ridiculous scientific names. To Joanna Wiśniowska, for answering my legal queries. To Andrew Steele, for inspiration, support, showing me the ropes, and being a trailblazer.

To Jenny Smith, my outstanding illustrator—for not only doing a magnificent job of capturing the protagonists of this book, but also for tirelessly dealing with my comments and suggestions, both the super-nerdy ones (on curvatures of beaks, shapes of ears, etc), and the, presumably much more frustrating, unspecific ones ("this bat-eared fox looks too proportional, please make it uglier"; "this wombat needs to look stupider, it has the air of an intellectual"). Jenny was all I could have hoped for, and more.

To my editor Lindsay Davies from Wildfire—for detailed and valuable feedback, which tended to be so positive and encouraging that at times I was suspicious about it all being some sort of a hoax. It seems that the job of an editor is as much that of a life coach as it is of a wordsmith, and Lindsay did both wonderfully. She was to this project what the tail is to the flying fish, giving it direction, momentum, and lift.

To the rest of the Wildfire team, especially Alex Clarke, and the copyeditor Lorraine Jerram, as well as the team at Smithsonian Books. To my agent Piotr Wawrzeńczyk and the Book/lab agency.

To my dad for feedback (naval-related and otherwise), and for bouts of extensive childcare; to my mom for unwavering enthusiasm and support, and for sending Dad over from Poland to do the aforementioned bouts of extensive childcare.

To my in-laws for forwarding news clippings of various cool Aussie animals (I'm sorry I couldn't include all of them!).

To my daughter Flora, for pitching ideas and gauging audience responses by sharing animal stories with her nursery group, and, later, school class. I can only apologise to her teachers.

To my daughter Antonia, for being a very relaxed baby, a good sleeper, and for patiently staring at my armpit as I was typing over her head when she was awake.

Finally, to my supportive and self-sacrificing husband Ian, who had to hear about all the weird animals ("could you please, for once, just wait until I am done with dinner?"), who read the chapters, laughed in the right moments, asked difficult questions, and was nothing short of amazing throughout the process.

Thank you!

Index

reptiles
birds
mammals
TARDIGRADA
amphibians
hagfish
bony fish
tunicates
cartilaginous fish
CHORI
PORIFERA
ECHINODERMA
CTENOPHORA